Structured Analysis
and Design
of Information Systems

A. Ziya Aktas

 Prentice-Hall International, Inc.

0-13-854571-5

Printed in the United States of America

10 9 8 7 6 5 4 3

ISBN 0-13-854571-5

Prentice-Hall International (UK) Limited, *London*
Prentice-Hall of Australia Pty. Limited, *Sydney*
Prentice-Hall Canada Inc., *Toronto*
Prentice Hall Hispanoamericana, S.A., *Mexico*
Prentice-Hall of India Private Limited, *New Delhi*
Prentice-Hall of Japan, Inc., *Tokyo*
Prentice-Hall of Southeast Asia Pte. Ltd., *Singapore*
Editora Prentice-Hall do Brasil, Ltda., *Rio de Janeiro*
Prentice-Hall, *Englewood Cliffs, New Jersey*

Contents

Foreword

By enabling us to gather a wealth of information on myriad subjects at an ever increasing rate, computers have ushered in an "information explosion." As a result we are gaining a new appreciation of information itself as an organizational resource. Our appreciation calls for even greater reserves of information and, accordingly, more efficient and effective ways to analyze and design successful information systems.

In this highly readable text, Dr. A. Ziya Aktas explains how structured system development methodologies surpass the capabilities of the classical approach in expanding the efficiency of information systems. Structure, he argues, is essential because it imposes order and thus improves the comprehensibility of complex systems. The structured approach presented herein is a combination of common sense and the rigorous application of some very useful tools.

Dr. Aktas writes as an accomplished teacher and researcher. He is currently Chairman of the Department of Computer Engineering at Middle East Technical University (METU) at Ankara, Turkey. During his fifteen years of teaching, Dr. Aktas has published several books and many papers in prestigious international journals. He received his B.S. and M.S. degrees in civil engineering from METU and his Ph.D. in civil engineering from Lehigh University.

While taking a two-year leave from METU, Dr. Aktas served on the faculty of the Computer Technology Department of Purdue University School of Engineering at Indianapolis, part of Indiana University–Purdue University at Indianapolis (IUPUI). During that time our faculty and students were fortunate to be introduced to many of the ideas and techniques presented in this text. We have witnessed their proven effectiveness.

I believe this book reflects Dr. Aktas' dedication to excellence. Both students and practitioners will find the text invites careful reading. Most assuredly the book will provide the reader with an increased awareness of the advantages and the necessity of the structured approach. If offers timely information regarding the tools and methodologies employed in the approach as well. Overall, it is an important contribution to the study of systems analysis and design.

R. Bruce Renda
Dean
Purdue University
School of Engineering and Technology
at Indianapolis

Preface

Both technical and nontechnical media have recently emphasized the growing need for systems analysts in the information systems field. The crucial point stressed repeatedly is that the number of available systems analysts is quite below the number that is demanded. A logical step in solving the problem would be to include some courses in systems analysis and design in the curricula of computer science, computer technology, information systems/science, computer engineering, business, or similar programs. This is in fact what is being done in many schools in the United States and Europe.

It has been observed, however, that teaching a systems analysis and design course is quite difficult. Perhaps the major limitation of conventional systems analysis and design courses is the lack of awareness of how to apply engineering tools and methods. If we begin by defining a structured approach as an engineering approach for information systems development, or as a well-defined and standard set of methods and tools for problem solving, we can then go on to teach how to use the structured tools and methods/methodologies in such courses.

The major objectives of this book are threefold: to present material that will clearly convey what are the available tools and methodologies for structured analysis and design of information systems, to compare these tools and methodologies, and to give examples of their applications. Hence it can be used either as a textbook or as a reference by the practicing systems analyst in a business or nonbusiness organization. A quite thorough review of recent publications in the field and a brief discussion of future research directions in the information systems development area are other contributions of the book. Some exercises and examples are provided within each chapter to illustrate the material, and a short case study is provided in the appendix to show the application of some commonly used methodologies such as SD (Structured Design), W/O (Warnier/Orr), and JSD (Jackson System Design). Relevant references will be given at the end of each chapter.

Some of the outstanding features of the book include:

- A discussion of classical approach vs. the structured approach in Part I should help clarify some often confusing terms and concepts.
- A practical life cycle for an information system is proposed.
- Objectives of the structured approach are elaborated.
- A thorough discussion of the available tools and methodologies of the structured approach reveals the entire spectrum of possibilities and the differences between tools and methodologies.
- The material is up-to-date and contains applications of various methodologies to the same or similar problems to facilitate comparison.
- Measurement of software properties using quantifiable metrics has been a relatively recent topic in software engineering area.
- A short chapter of the book is devoted to a relatively important and promising research topic now being directed at applying concepts to the measurement of properties of information systems.

It would be very appropriate to use this text for a senior or a graduate course that is taught after a classical systems analysis and design course. The book is primarily intended to be used as a text book for any course similar to CIS-5 Structured Systems Analysis and Design of the DPMA Curriculum. It can, of course, also be used by professional systems analysts and programmers as well.

I am indebted to many people who aided in the preparation of the book. It is now my pleasure to acknowledge some of these very fine people who encouraged, helped, and gave me the strength that I needed to finish the manuscript. First I would like to acknowledge B. Elderd of Reston/Prentice-Hall, Inc. Her continuous interest and care gave me the courage to start and to pursue my efforts in manuscript preparation. I would like to thank her, Reston/Prentice-Hall, and the discipline editors, J. Sulzycki and T. Buchholz, and editorial associate J. Kinzer for their care, interest, and support in the preparation and review of the manuscript. Actually, executive editor J. F. Fegen, and production editors C. Atkins and N. Menges, of Prentice-Hall made editing and publication of this book possible. It is my real pleasure to express my gratitude and sincere thanks to them and to all other members of Prentice-Hall who contributed to the editing, production, and manufacturing. Many thanks are due various anonymous reviewers who contributed greatly by their comments, criticisms, and encouragements. It is another pleasure to express my thanks to B. Renda and R. Crozier of Purdue University School of

Engineering and Technology at Indianapolis, the dean and the department chairman, respectively, for their support and encouragement during the manuscript preparation. How can I forget the late J. Williams? I owe him and our dear friend E. Solinski many thanks for the very valuable discussions that we held together and also for their encouragement in my work. A few chapters of the book were drafted by J. Keely and the rest were drafted and typed by G. Oskay. It was later retyped a few times by A. Yardim. It is difficult for me to express my gratitute and thanks to her and others. My students in Indianapolis and in Ankara contributed a lot by their care, interest, questions, and comments during and after my lectures on structured approaches. I thank them all. I also thank Z. Aykanat and H. Sarrafzadeh for their assistance in preparing the manuscript. Last but not least, I want to express my sincere thanks to my family: my wife, our boys, and our parents. Without their patience, understanding, and sacrifice I would never have been able to finish this work.

A. Ziya Aktas
METU—Ankara

Chapter 1

Fundamental Concepts

1.1 SYSTEMS CONCEPT AND INFORMATION SYSTEMS

"System" is a term that is commonly used in many disciplines. Its use increased remarkably after the late fifties especially due to the rapid growth of electronic data processing (EDP) activities and to the appearance of some interdisciplinary studies such as cybernetics and biophysics. Social, educational, industrial, business, and engineering systems are only a few examples of systems in today's world. Obviously there exist many definitions of system.

Although the word "system" is used in many seemingly unrelated areas, all systems have some common properties: They have elements, environments, interaction between their elements and with the environment, and most important of all, they have goals to be fulfilled. Hence, one may define a system as an organized collection of people, machines, procedures, documents, data, or any other entities such that they interact with each other as well as with the environment to reach a predefined goal. A "subsystem" is a system which is an element of a larger system. The large system is called a "super system" or "supra system."

Very similar to the word "system," the word "information" is also a very much used and confused word. It is used by EDP people, by communications people, by librarians, and by many others. "Data" is another word that is used with information or even as its synonym. Yet a third word, "knowledge," is closely related to the other two. In fact fifth-generation computer systems are being referred to as knowledge-information processing systems (TL 82). An attempt to clear that confusion around the terms data/information/knowledge needs to refer to another term: message. A message is a group of characters that is stored, processed, and transmitted in the information system of an organization. In other words, information in an organization is stored, processed, and transmitted as messages. The content of messages that flow through the information system of an organization has different levels of meaning depending on whether the messages carry data, information, or knowledge. As pointed out by Taggart (Tag 80), we recognize three levels of meaning for messages included in an information system: data, information, and knowledge (Figure 1.1).

- knowledge
- information
- data   meaning value   **Figure 1.1**   Meaning levels of messages

Data are groups of characters recognized as having the lowest level meaning. They are raw facts and opinions. *Information* has more meaning than data in that it is useful in a present decision situation. *Knowledge* has the highest level of meaning because it represents information that can be potentially useful in future decision situations.

As a comparison of these three terms, consider the following illustration:

Data: Employees have submitted their vacation requests.
Information: Summer is the vacation season, but we have already received production orders for summer.
Knowledge: Employee vacation dates should be arranged to handle the summer production properly.

There is still difficulty in distinguishing data from information or knowledge, however, because a certain data element may be information to a user at one time and knowledge to the same user at a different time or place. We, therefore, use the term "information" in the text as a general term, without differentiating the meaning level of the messages.

The information system of an organization may be defined as a system that serves to provide information within the organization when and where it is needed at any managerial level. Such a system must take the information received and store, retrieve, transform, process, and communicate it using the computer system or some other means. Bryce (BrM 83) states: "Equating information systems with computer systems is a misconception born of decades of preoccupation with technology." He defines an information system as a logically interrelated set of business processes that accomplish organizational goals.

As noted by Taggart (Tag 80), information in an information system environment has the following common requirements:

- It must be understood by its recipient in the proper frame of reference.
- It must be relevant to a current need in the decision-making process.

- It must have a surprise value, that is, what is already known should not be presented.
- It must lead its users to make a decision. A decision could be to take no action.

Brookes et al. (BGJL 82) stated that for an information system to yield information with the above characteristics, it must have some attributes. We may rearrange them as follows:

- Effective processing of information. This refers to proper editing of input data and efficient utilization of hardware and software.
- Effective management of information. Stressed are care in file management operations and in security and integrity of the existing data.
- Flexibility. The information system should be flexible enough to handle a variety of operations.
- User satisfaction. Of prime importance are user understanding of and satisfaction with the information system.

Any action in an organization depends on the result of a decision-making process at the proper managerial level of that organization. Obviously, then, the right information, available at the right time, is vital for the process of optimal decision making. In some decision-making/action processes the available data can be used directly. In most cases, however, data are grouped and summarized in various forms and finally transformed to become information to be used in decision making. The result of action(s) following a decision-making process can also be used for another decision and action activity (i.e. feedback process in systems terminology). The relationship between data and action is illustrated in Figure 1.2 and draws on that given by Konsynski (Ko 80).

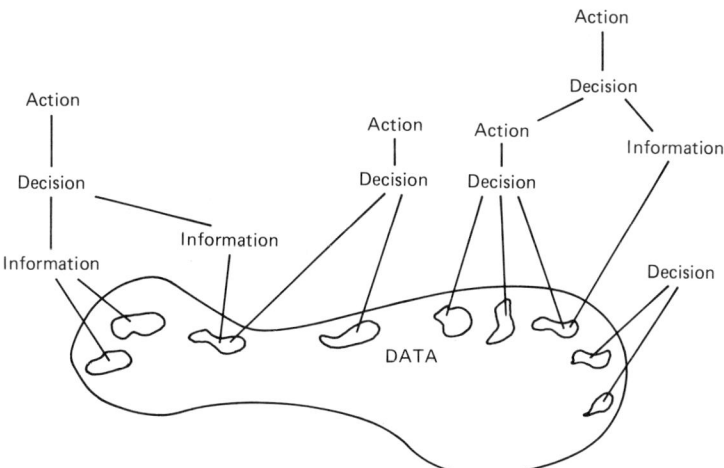

Figure 1.2 Data/action relationship in an organization

1.2 SUBSYSTEMS OF AN INFORMATION SYSTEM

The information system of any organization contains information related to three basic types of operations, namely, transaction processing, control, and strategic planning. Following Davis (DaG 74), one could group them into two as operating level and management level activities as in Figure 1.3.

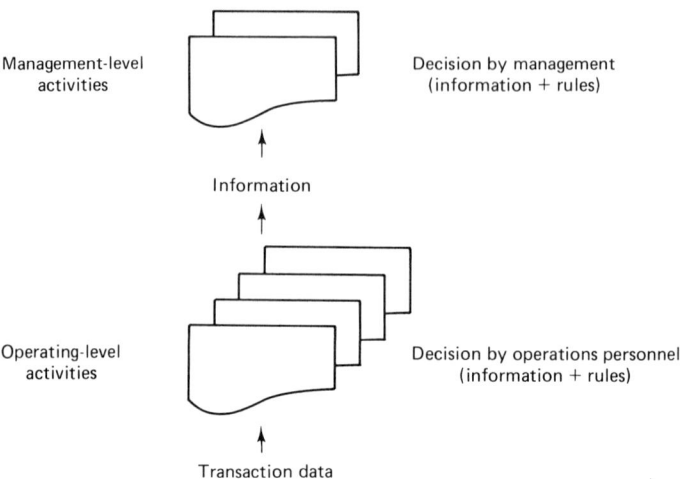

Management-level activities

Decision by management (information + rules)

Information

Operating-level activities

Decision by operations personnel (information + rules)

Transaction data

Figure 1.3   Operating level and management level activities

It is almost conventional now to represent management level activities as a triangle and to base it on a rectangle as a symbol of operating level activities (Figure 1.4).

Let us first consider the management level activities. In strategic planning, the executive or top management of the organization decides on the objectives of the organization, on the resources to be used to attain these objectives, and on the policies that are to govern the acquisition, use, and disposition of these resources. Activities have a long time range—one to ten or more years. As seen in Figure 1.4, the control function has both management and operational components. In *management control*, middle-level managers assure that resources are obtained and used effectively and efficiently to accomplish the organization's objectives. Activities have year-to-year or monthly time range. In *operational control*, supervisory management assures that specific tasks are carried out effectively and efficiently. Activities have day-to-day or weekly time range (e.g., IBM 78).

The rectangular block under the management triangle in Figure 1.4 is used to represent *transaction processing*, which means the daily routine business operations of the organization.

When we consider the relationship between the information system of an organization and its activities, common practice has been to define two subsystems

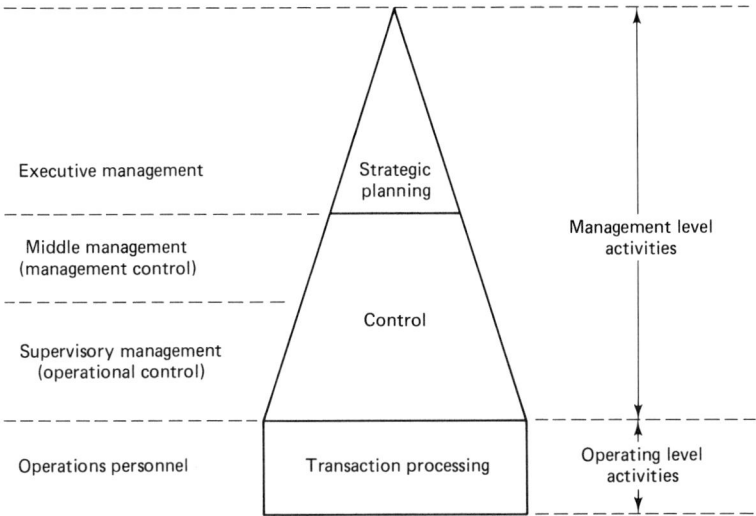

Figure 1.4   Information-related activities of organizations

as MIS (Management Information System) and OIS (Operations Information System) as in Figure 1.5. Comparing Figures 1.4 and 1.5, one concludes that OIS is the information subsystem relevant to transaction processing of the organization, and MIS is the information subsystem relevant to managerial decisions for control and strategic planning purposes.

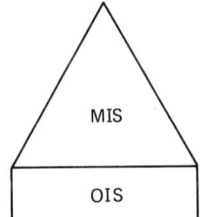

Figure 1.5   Information subsystems of an organization

1.3 AN EXAMPLE

Consider a small manufacturing company, Central Circuits Co., whose primary business objective is to produce custom-engineered printed circuit cards for larger electronic component manufacturers. When we review the types of activities of this company, we can combine Figures 1.2, 1.3, and 1.4 into a schema as shown in Figure 1.6.

In that figure data/information/knowledge (decision) relations are illustrated

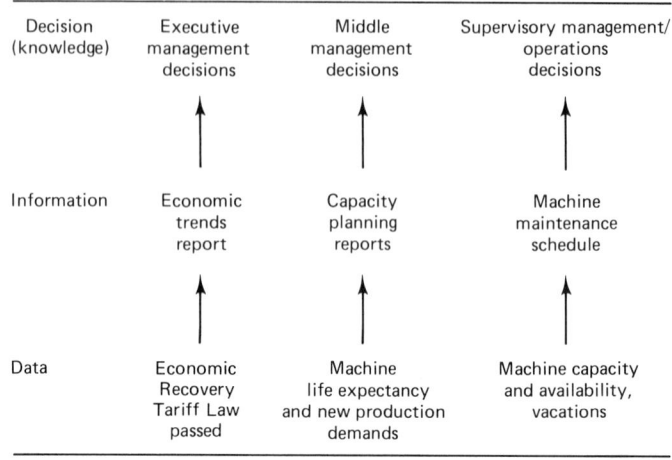

Figure 1.6   Example of an information system

using some examples of decisions that would be made by executive management, middle management, and supervisory management.

1.4 EDP/MIS/DSS

We have conceptualized the information system of an organization as consisting of two subsystems: MIS (Management Information System) and OIS (Operations Information System). Quite often the OIS of an organization is referred to as EDP (electronic data processing) if the information flow is supported by computers. In addition, the tendency now is to break down the MIS subsystem of an information system into MIS and DSS (Decision Support Systems). Various papers and books have appeared to refine the DSS idea (e.g., BHW 81, SC 82, Th 83) and to define the boundaries between EDP, MIS, and DSS (e.g., Di 81, RB 82).

As stated by Raho and Belohlav (RB 82), the flood of information presents both a threat and an opportunity for the manager. The information overload that can inundate a manager at any managerial level of the organization may be a threat. More information, however, is an opportunity for the manager to make better informed decisions. The effective collection, storage, transformation, and communication of information in an organization is a function of EDP, MIS, and DSS. Understanding the differences between EDP, MIS, and DSS will, therefore, help managers to cope with the dynamic information systems they are living with.

Referring to Raho and Belohlav (RB 82) again, we may state that data are of primary interest in an EDP system which functions at an operational level within the organization. The internal output of an EDP system consists mainly of factual reports and summary reports.

In an MIS, information is data which has been processed to become meaningful to a variety of potential users, primarily middle and upper management. In contrast to the EDP system, an MIS views not just the transformation of data but how it can be turned into useful organizational information. The major output of an MIS consists of standardized reports and interrogative reporting. An MIS is designed from an organizational perspective rather than from a business transaction perspective.

In DSS, the emphasis is on decision making at all levels of management in the organization. DSS is a tool to help managers in their decision-making activities by providing them with the necessary information. The information produced by a DSS consists of interactive/iterative reports and unstructured reports oriented to the individual manager. An important distinction between DSS on the one hand and EDP and MIS on the other has to do with development of the system structure. In EDP and MIS, the systems designers develop the system structure, whereas in DSS the manager not only provides the input but he/she defines the structure as well. Another definition of DSS is given as: "DSS means marrying analytical tools with computer technology and putting it out in executive users' offices" (Th 83).

1.5 OTHER ASPECTS OF INFORMATION

Information is not only a resource but an asset for an organization, empowering it to produce changes in its environment. "Today, businesses are recognizing data more and more as a resource that is as important as personnel, cash, facilities, or materials. They see the need to consolidate the key data files and make information available not just to individual functions as departments but throughout the business, in order for management to gain an overall view of the business and be able to make multifunctional decisions" (IBM 81). It is also predicted that "the organizations that will excel in the 1980s will be those that manage information as a major resource" (BrM 83). In the proceedings of a conference, Oren (Ore 81) quotes a comment to the effect that "a number of countries in the West are now passing into a post-industrial phase of society where the strategic resource is knowledge as opposed to raw materials or financial capital which were the strategic resources in pre-industrial and industrial societies, repectively." Defining knowledge as high-level information, one sees the significance of information in the technical development of countries.

The data communications system of an organization is another topic that is closely related to the information system of that organization. The transmission of data/information from one point to another in the organization by any means—for example, telephone lines or coaxial cables—is data communications, and it seems that more and more, data communications and information systems of organizations are becoming inseparable. Hampel (Ha 81) noted the convergence of information processing and data communications systems of organization and defined the eighties as a "decade of information" (Figure 1.7). As if to prove the validity of Figure 1.7, in 1982 AT&T restructured itself to enter the computer area. In the same year IBM emerged from an antitrust suit unharmed and announced its deci-

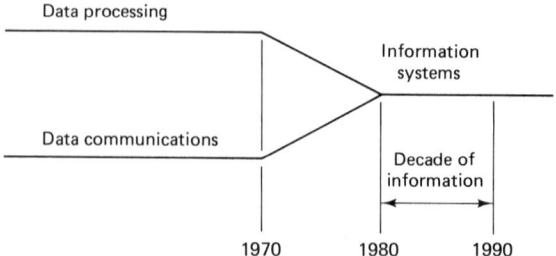

Figure 1.7 Information systems in the 1980s

sion to become more deeply involved with data communications. Taggart predicted these developments in AT&T and IBM as early as 1980 and described the situation as a "battle of giants" (Tag 80, p. 28). Clearly every year in this decade is witnessing the fact that information is becoming a key issue. "Information engineering" degree programs in some European countries indicate the significance of recent developments in this area.

The fast growing use of databases (DB) and database management systems (DBMS) is also closely related to data processing/data communications and information systems, and the proper application of these systems definitely improves the value of information in organizations. One of the major advantages of DB applications is data integrity or data correctness. A DB also improves the value of information by minimizing data redundancy. Another major advantage of DB is that the same set of data is available for different applications.

Looking again at our hypothetical small company, Central Circuits Company, we may describe its DB as in Figure 1.8.

As can be seen from the figure, the major groups of data are raw material data, machine capacity data, process data, demand data, product definition data, and personnel resources data. The capacity planning system, work order control system, and economics and market analysis system are the main systems utilizing the DB. In the figure it is intended to describe the data as the content of a single DB. For a large organization, however, each or some combination of these data form an individual DB.

Another important factor related to the success of information systems as they impact on decision-making activities is the reliability of the information in the database system. Zakay (Zak 82) considers the reliability of information as a potential threat to the acceptability of decision support systems. He suggests three main potential sources as the major causes for an unreliable database: (1) information that is held back by users and not entered into the system, (2) biased information entered into the system, and (3) incorrect updating of the database.

The developments in "knowledge-based expert systems," or "knowledge system" for short, are expected to lead over time to the construction of extremely valuable "knowledge bases," a basic unit of the new discipline of "knowledge engineering." It is predicted that knowledge systems will increase individual and

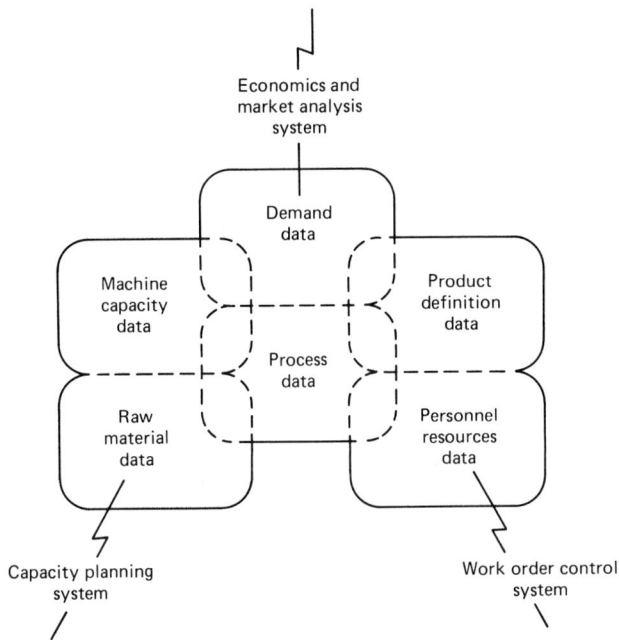

Figure 1.8   Database of Central Circuits Company

social potential by preserving know-how, by distributing knowledge more effectively, and by improving the performance of tasks that require expertise (HR 84). It was argued by Keen (Kee 81) that information systems development is an intensely political as well as a technical process and that organizational mechanisms are needed to provide MIS managers with authority and resources for negotiation. Finally, it was pointed out that MIS can be improved by understanding the behavioral processes by which humans process information and make choices (BT 82).

In a recent survey conducted by Computerworld about the problems of data processing (DP) managers, applications backlog was identified as their biggest headache (Com 83). The next most pressing problems were found to be keeping up with technology and budget cuts. Obviously rapid advances in hardware/software components of computers are rapidly increasing the number of computer-based applications. Thus the demand for information-processing professionals with problem-solving capabilities continues to grow. Trying to meet that demand necessitates revising the curricula of existing academic programs and encourages the proposal of new curricula.

SUMMARY

"System" is a term that is common to many disciplines. All systems have some common properties: They have elements, environments, interaction between their elements and with the environment, and goals to be fulfilled. A system may be defined as an organized collection of people, machines, procedures, documents, data, or any other entities, such that they interact with each other as well as with environment to reach a predefined goal.

Information, data, and knowledge are closely related terms. They may be defined as the meaning content levels of messages flowing in an information system.

The information system of an organization may be defined as a system that serves to provide information in the organization when and where it is needed at any managerial level. Information in an information system must be understood by its recipient in the proper frame of reference; it must be relevant to a current decision-making process, it must have a surprise value, and it must lead its users into action. Effective processing and management of information, flexibility, and user satisfaction are some of the desirable attributes of an information system.

Any action in an organization depends on the result of a decision-making process at the proper managerial level of that organization. The right information, available at the right time will, therefore, be vital in the decision/action activities of managers.

Transaction processing, control, and strategic planning are those operations of an organization for which information is kept in the information system. Information about these operations may be defined in two subsystems as MIS (Management Information System) and OIS (Operations Information System). Recently computer-based and information-related activities of an organization are grouped under EDP, MIS, and DSS.

The value of information in organizations is a critical issue. It is generally accepted now that information is an asset as well as a resource for an organization.

Data communications is another topic that is highly related to the information system of organizations.

Databases (DB) and database management systems (DBMS) are closely related to data processing/data communications and information systems, and their proper applications improve the value of information in organizations.

Reliability of information in the databases, behavioral aspects of information systems, and problem-solving capability of information systems professionals are some of the recent issues relevant to information and information systems.

EXERCISES

1. Give three examples of systems.
2. What are the common properties of systems?
3. Define a system.
4. What is a message?
5. Define data, information, and knowledge.
6. What is the information system of an organization?

7. What are the common requirements for information in an information system environment?
8. What are some attributes of an information system?
9. What is the significance of information in decision-making/action process?
10. What are the basic types of operations for which information is kept in the information system of an organization?
11. What is strategic planning?
12. What are the control activities of an organization?
13. What is transaction processing in an organization?
14. What are the subsystems of an information system?
15. Comment on the differences between EDP, MIS, and DSS.
16. Discuss the value of information for organizations.
17. What is the relationship between the data communications and information systems of an organization?
18. Why do we call the eighties a "decade of information"?
19. What are the major advantages of DB applications?
20. What is the effect of reliability on information systems?
21. What are the major points in the behavioral aspects of information systems?

SELECTED REFERENCES

(BGJI 82)   Brookes, C. H. P. et al. *Information Systems Design*. Prentice-Hall, 1982.

(BHW 81)   Bonczek, R. H., C. W. Holsapple, and A. B. Whinston. *Foundations of Decision Support Systems*. Academic Press, 1981.

(BT 82)   Benbasat, I., and R. N. Taylor. "Behavioral Aspects of Information Processing for the Design of MIS," *IEEE Transactions on Systems, Man, and Cybernetics*, Vol. SMC-12, No. 4 (July/August 1982), pp. 440–450.

(BrM 83)   Bryce, M. "Information Resource Mismanagement," *Infosystems*, No. 2, February 1983, pp. 89–92.

(CCEH 81)   Cotterman, W. W. et al. *Systems Analysis and Design*, North Holland, 1981.

(COM 83)   *Computerworld*, Weekly Newspaper, CW Communications/Inc., April 11, 1983, p. 1.

(DaG 74)   Davis, G. B. *Management Information Systems: Conceptual Foundations, Structure, and Development*. McGraw Hill, 1974.

(Di 81)   Dickson, G. W., "MIS: Evolution and Status," in *Advances in Computers*, ed. M. C. Yovits. Academic Press, 1981, pp. 1–37.

(Ha 81)   Hampel, W. E., "Fact Retrieval in the 1980's," NATO/AGARD Conference Preprint No. 304, August 1981, pp. 6: 1–36.

(HR 84)   Hayes-Roth, F. "Knowledge Based Expert Systems," *IEEE Computer*, October 1984, pp. 263–273.

(IBM 78)   IBM. *Business Systems Planning*, GE 20-0630-0, November 1978.

(IBM 81)   IBM. *Information Systems Planning Guide*, GE20-0527-3, July 1981.

(Kee 81)   Keen, P. G. W., "Information Systems and Organizational Change," *Comm. ACM*, Vol. 24, No. 1 (January 1981), 24–33.

(Ko 80)   Konsynski, B. "Data Base Driven System Design," in *Systems Analysis and Design*, ed. W. W. Cotterman et al. North Holland, 1981, pp. 251–278.

(Ore 81)   Oren, T. I. "Foundations for an Information Technology," *Proc. of 1981 Winter Simulation Conf.*, ed. T. I. Oren et al., ACM/IEEE, 1981, pp. 201–208.

(RB 82)   Raho, L. E., and J. A. Belohlav, "Discriminating Characteristics of EDP, MIS and DSS Information Interface," *Data Management*, December 1982, pp. 18–20.

(SC 82)   Sprague, R. H., and E. D. Carlson. *Building Effective Decision Support Systems*. Prentice-Hall, 1982.

(Tag 80)   Taggart, W. *Information Systems*. Allyn and Bacon, 1980.

(Th 83)   Thiel, C. T. "DSS Means Computer-Aided Management," *Infosystems*, No. 3, 1983, pp. 38–44.

(TL 82)   Treleaven, P. C., and I. G. Lima, "Japan's Fifth-Generation Computer Systems," *IEEE Computer*, August 1982, pp. 79–88.

(Yov 81)   Yovits, M. C. *Advances in Computers*. Academic Press, 1981.

(Zak 82)   Zakay, D. "Reliability of Information as a Potential Threat to the Acceptability of DSS," in *IEEE Transactions on Systems, Man, and Cybernetics*, Vol. SMC-12, Nol. 4 (July/August 1982), 518–520.

Chapter 2

Information Systems Life Cycles

2.1 INTRODUCTION

In the mid-sixties, there occurred a number of large, costly, and embarrassing failures in EDP applications for large systems, in large part because of poor or nonexistent system development techniques (e.g., Fr 79). Following these failures, an understanding of the significance of information systems development methodologies began to emerge in the late sixties and early seventies. Since then various methodology proposals have been made and their applications reported. Designers of almost all of these information systems development methodologies have had a common point: They have realized that, computer-based or not, any information systems development process is, or rather must be, like an engineering systems development process.

With a view to construction and operation of various types of buildings, power transmission lines, various machines, and chemical plants as examples of engineering system development, we can summarize the major phases in such a development process as

1. Planning
2. Analysis
3. Design
4. Implementation or construction
5. Maintenance

In the planning phase, the engineer gathers information about the problem and the requirements. He/she then sets criteria and constraints for a solution and generates a number of alternative solutions. In the analysis phase, the engineer tests the alternative solutions against the criteria and constraints. The analysis is a pivotal point in the entire development process. As noted by various authors (e.g., EJMN 79, In 78a), a great deal of engineering education deals with teaching engineers how to analyze. The laws of nature, the rules of economics, and common sense (often referred to as engineering judgment) are the main elements of analysis conducted by engineers. The next major phase in the engineering system development, namely design (or synthesis), may be defined as the creation of a new system in accordance with a preconceived plan from the analysis phase. The design or synthesis phase may also be defined as "the optimum solution to the sum of the true needs of a particular set of circumstances" or as "a creativity activity—it involves bringing into being something new and useful that has not existed previously" (In 78a, p. 8). The designed system is constructed and operated. Maintenance is performed on every system which is operational.

The "life cycle" of a system is a term used to describe the major phases and their steps in its development process. Clearly then, life cycles of engineering systems and information systems should be the same or similar and the same general principles should be held valid in information systems development. In the following sections this point will be seen more clearly by comparing the life cycles of information and engineering systems; later a life cycle for information systems will be given.

2.2 INFORMATION SYSTEMS LIFE CYCLE
VS. ENGINEERING SYSTEMS LIFE CYCLE

As noted earlier, to be able to see the similarity between development processes of information systems and engineering systems, one should compare their life cycles. In Table 2.1 we tabulate the life cycles for four information systems between 1960 and 1983 (Op 60, Kel 70, BBA 80, and DaW 83). Examining the life cycles for information systems that have been proposed by different authors at various times during the last 20 years, one notes their similarity. But the most crucial point is that information systems life cycles are very close to those followed by engineering systems; that is, planning, analysis, design, implementation, and maintenance are also the major phases for developing information systems. This is not a coincidence; once more it should be stressed that an information system development process is an engineering process and as such has to follow the same steps and obey the same general principles, as we shall elaborate in the next chapters. One should also refer to the term "software engineering" to recall the engineering nature of the software development process, which is a subsystem of an information system.

Another interesting conclusion from the above information systems life cycle

TABLE 2.1 Information Systems Life Cycles for Two Decades

Steps	1960 (Op 60)	1970 (Kel 70)	1980 (BBA 80)	1983 (DaW 83)
1	Analyze the present system	Scope definition	Initial investigation	Problem definition
2	Develop a conceptual model	Survey study	Feasibility study	Feasibility study
3	Test the model	Data collection and analysis	Operations and system analysis	Analysis
4	Pilot installation of the new system	System design	User requirements	System design
5	Full installation of the new system	Implementation planning	Technical support approach	Detailed design
6		Development	Conceptual design and package review	Implementation
7		Testing	Alternatives evaluation and planning	Maintenance
8		Cutover	Systems technical specifications	
9		Maintenance	Technical support development	
10			Applications specifications	
11			Applications programming and testing	
12			User procedures and controls	
13			User training	
14			Implementation planning	
15			Conversion planning	
16			Systems test	
17			Conversion and phased implementation	
18			Refinement and tuning	
19			Post-implementation review	

Groupings for 1970 (Kel 70):
- Steps 1–2: System Survey
- Steps 3–5: System Analysis and Design
- Steps 6–9: System Development

Groupings for 1980 (BBA 80):
- Steps 1–2: Systems Planning
- Steps 3–7: Systems Requirements
- Steps 8–16: Systems Development
- Steps 17–19: Systems Implementation

study is the fact that although information systems life cycles have been known to consist of more or less the same steps for almost 20 years, the naming and even the adherence to these steps has not sufficed to develop successful information systems. As we shall discuss further in the next chapter, there has been something missing in information systems development efforts. The missing piece is the fact that in any engineering system development there are some tools and methodologies to be used in parallel with the system life cycle. Failure in determination of user requirements and in user participation for system development has been another critical reason for unsuccessful information systems, in addition to hard sell from computer manufacturers, inadequate staff, unrealistic deadlines, ignorant management, and some other common causes.

2.3 A PROPOSAL FOR AN INFORMATION SYSTEMS LIFE CYCLE

Table 2.2, graphically represented in Figure 2.1, proposes a new life cycle for information systems.[1] A common misinterpretation of tables like Table 2.2 is the impression that the whole process is linear: It looks as if all the phases and steps follow each other in a sequential manner. This is, however, not the reality. All the phases and their steps of the development process have an iterative nature; that is,

---

[1]Table A.1, given in Appendix A, includes the key considerations and end products of the development steps depicted in Table 2.2.

TABLE 2.2   **Information Systems Life Cycle**

 I. Planning
    1.1. Request for a system study
    1.2. Initial investigation
    1.3. Feasibility study
 II. Analysis
    2.1. Redefine the problem
    2.2. Understand the existing system
    2.3. Determine user requirements and constraints on a new system
    2.4. Logical model of the recommended solution (conceptual, logical, or architectural design)
         or functional specifications
 III. Physical Design
    3.1. System design (or general design or system specifications)
    3.2. Detailed design (or specific design)
 IV. Implementation or Construction
    4.1. System building
    4.2. Testing
    4.3. Installation/conversion
    4.4. Operations (refinement/tuning)
    4.5. Post-implementation review
 V. Maintenance
    5.1. Maintenance and enhancements

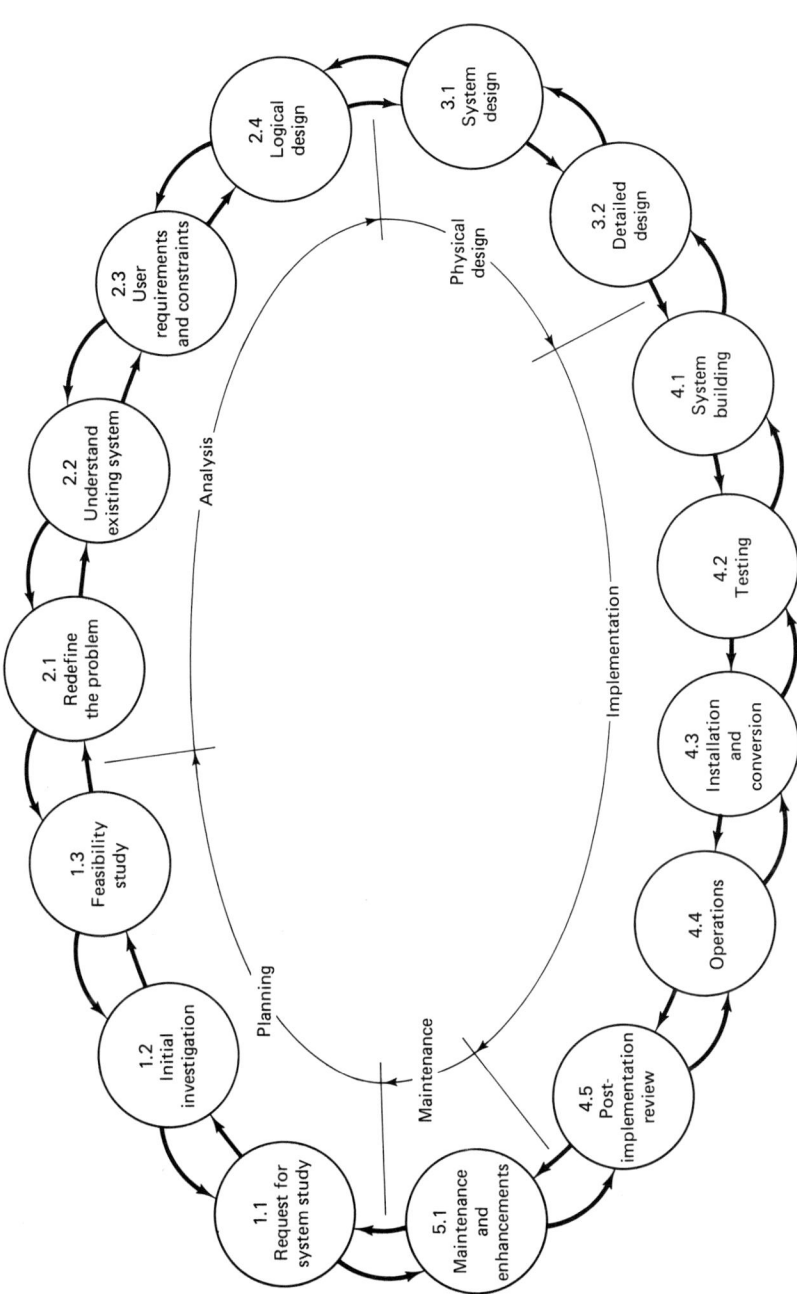

Figure 2.1 Information systems life cycle

work on a step or a phase often requires to go back to the previous step(s) or even phase(s), and whatever has been done up to that point may need to be completely revised. Maintenance of a system usually requires the repetition of the whole development process starting with the planning phase. In order to depict the cyclic nature of the information system development process, we have used an elliptical shape in Figure 2.1. The reverse directed arrows between the steps are used to indicate the iteration of some steps and phases, that is, to imply the nonlinear character of the development process.

A number of methodologies for information systems development consisted simply of following steps similar to those given in Table 2.2 or Figure 2.1. Such a methodology or approach is commonly known as "classical systems approach" or "classical (traditional or conventional) analysis and design of information systems."

Although there may be some variations among individual steps, the "classical systems approach" states that following the steps of an information system life cycle will yield a successful information system. Unfortunately, however, this practice has not been sufficient to yield a successful information system; in addition to naming the stages of a system life cycle, one has to have some standard tools and techniques to develop that system. Beginning in the early seventies, some tools and methodologies were gathered under the name of "structured approach" or "structured systems development methodologies" or "structured analysis and design methodologies." They basically provided the systems analyst with additional tools and techniques, besides the idea of an information system life cycle. The need for such an approach will be discussed further in the next chapter.

We should mention here that although there have been continuous efforts to develop and reevaluate current information systems, the emphasis of the classical approach has been on technical personnel, not on the user. Only recently has it been realized that user understanding and user support for a system being developed is vital for the success of the final system. One of the main contributions of structured methodologies for information systems development is thus user participation and consequently user commitment. Of course, this commitment ideally begins with the executive management of the user organization.

SUMMARY

The major phases and steps of a system development process are commonly known as a "life cycle." Life cycles of engineering systems and information systems are very similar, if not the same. The classical approach to information system development emphasizes the use of a life cycle and documentation for system development. Since the early 1970s, however, there is a new approach called the structured approach, which provides the systems analyst with some tools and methodologies in addition to the life cycle concept to develop a successful information system.

EXERCISES

1. What are the major phases of an engineering system development process?
2. Compare the information systems life cycles for the period 1960 to 1983. What are the similarities and differences?
3. Is information system development an engineering process? Why?
4. What do we mean by a cycle and by the iterative nature of an information system life cycle?
5. What is the basic idea behind the classical approach to information systems development?
6. What is the basic difference between the ''classical approach'' and the ''structured approach'' to information systems development?

SELECTED REFERENCES

(BBA 80)   Biggs, C. L., E. G. Birks, and W. Atkins. *Managing the Systems Development*. Prentice-Hall, 1980.

(BGJL 82)   Brookes, C. H. P., P. J. Grouse, D. R. Jeffery, and M. J. Lawrence. *Information Systems Design*. Prentice-Hall, 1982.

(DaW 83)   Davis, W. S. *Systems Analysis and Design*. Addison-Wesley, 1983.

(EJMN 79)   Eide, A. R., R. D. Jension, L. D. Mashaw, and L. L. Northrup. *Engineering Fundamentals and Problem Solving*. McGraw Hill, 1979.

(Fr 79)   Freeman, P. ''A Perspective on Requirements Analysis and Specification,'' Auerbach Publishers Inc., Portfolio No.: 32-04-01, 1979.

(In 78a)   Infotech State of the Art Report: Structured Analysis and Design, Vol. 1: Analysis and Bibliography, Infotech International Ltd., Maidenhead Berkshire, UK, 1978.

(Kel 70)   Kelly, J. F. *Computerized Management Information Systems*. MacMillan, 1970.

(Op 60)   Optner, S. L. *Systems Analysis for Business and Industrial Problem Solving*. Prentice-Hall, 1960.

Chapter 3

Classical Approach

3.1 INTRODUCTION

As noted earlier, the ''classical approach'' to information systems development consists of following steps similar to those given in Table 2.2. Because of the complicated nature of information systems, however, accepting and following the steps of the life cycle has not been enough to develop a successful information system. Remember, every engineering discipline—electrical, mechanical, structural, and so on—has its own methodology that has helped concretize system life cycle definitions. Unfortunately, until recently no one realized that following the life cycle alone would not yield successful information systems. Now, however, it is understood that the systems development approach requires some tools and techniques to make it successful. Such an approach is what we commonly call a ''structured approach'' or ''structured analysis and design of information systems.''

   In the following sections, the classical approach and its problems will be reviewed first. Recent trends in information systems life cycles will be discussed later.

3.2 PROBLEMS OF THE CLASSICAL APPROACH

In order to understand the need for a structured approach in information systems development, let's consider briefly the problems we have had using classical approach for information systems development:

3.2.1 Character of Classical Approach

As noted by McGowen and McHenry (McGMcH 80) the classical approach appeals to many because it is a contractual model. It is a familiar model embodied explicitly in contracts and institutionalized as the development and documentation standards in many organizations, more a managerial than a technical tool. Indeed the classical approach emphasizes documentation almost to the exclusion of development. Although documentation is a must in information systems development, the number of critical problems and failures in information systems has already shown that documentation alone is not enough for successful information system development. The development process itself must receive more emphasis than documentation, and documentation should be viewed simply as a by-product of a development process.

3.2.2 Hardware/Software Cost

Another reason for using a structured approach is the trend of software/hardware cost. Ramamoorthy et al. (RPTU 84) stated the following:

> Computer users first became aware of a software crisis 15 years ago. Software projects were being delivered far behind schedule, quality was poor, and maintenance was expensive. And as more complex software applications were found, programmers fell further behind the demand and their results were of poorer quality. The high demand and comparatively low productivity drove software costs up. In the US in 1980, software cost approximately $40 billion, or two percent of the gross national product. Dolotta estimates that by the year 1985 the cost of software will be approximately 8.5 percent of the GNP, while Steel points to 13 percent by 1990.

A relatively old prediction made by Boehm (Bo 76) is given as Figure 3.1. There are some arguments in the literature about the validity of Boehm's prediction; for

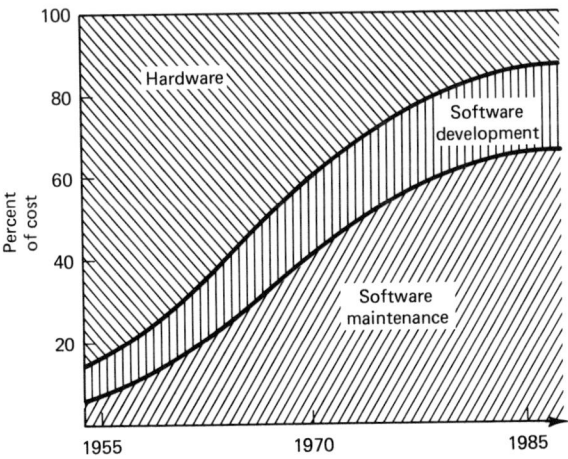

Figure 3.1   Computer-based information systems cost trends

example, an article by Cragon provided some numerical data to lower the ratio of software to hardware (Cr 82). It is clear, however, that the cost of software is continuously increasing while that per unit hardware is decreasing. Yet the demand for software is also increasing at an even faster rate. A few years ago Myers (MyW 78) pointed out that the need for software was increasing because access to inexpensive hardware brought with it many new applications requiring innovative software. The need is even greater now because of the microcomputer explosion that we are all witnessing. If we consider software to be an information subsystem, clearly the classical approach has failed to decrease these costs.

3.2.3 Maintenance Cost of Existing Information Systems

Another important reason for using the structured approach in information system development has to do with the maintenance cost of existing software and the information system as a whole (e.g., Ak 82, MyW 78, Ca 81). Normally corrective and/or enhancement maintenance of software accounts for about 70 percent of the total software cost. (Actually, two recent articles use the figure of 67 percent [CW 82, RPTU 84].) The overall cost of software and maintenance in the United States is estimated to be roughly 15 to 25 billion dollars per year. Although structured programming techniques have increased programmer productivity, that increase, which is estimated to be roughly 5 percent (MyW 78), lags far behind the rate of increase of total software cost. There is obviously a need to decrease the cost of software development, to increase reliability of the product and maintainability of the system, and to increase the productivity of the technical personnel involved in system development. Boehm (Bo 77) indicated that from two to four times the cost of development can be spent on system maintenance. Connor (Co 80) likened maintenance to the body of the cost iceberg (Figure 3.2).

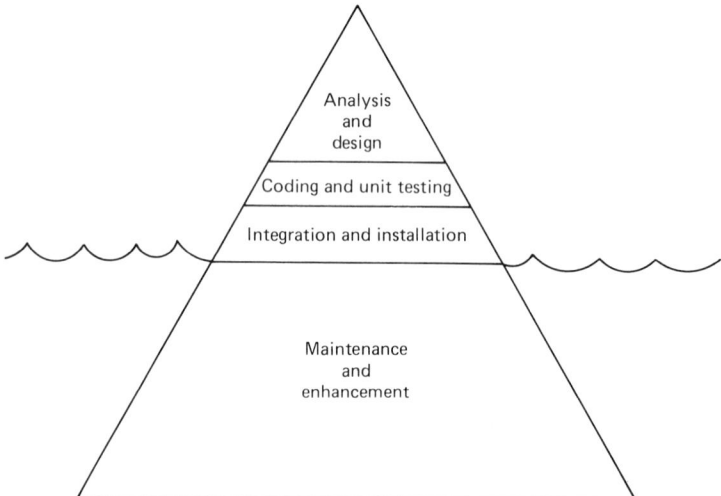

Figure 3.2   Computer-based information systems development and maintenance cost

3.2.4 User Requirements, Responsibility, and Involvement

The most critical shortcoming of the classical approach has been very limited reflection on user requirements in the completed information system. Ongoing communication with the user during development of the information system and after implementation is vital for the success of the project. The classical approach assumed that users know and state their requirements very clearly and correctly by the end of analysis phase—that is, before the physical design of the system. Experience has shown that in most cases users' requirements are not very clear and/or correct throughout almost the entire system development phase. As a result, system designers are now trying to involve users during the entire system development process and to have a development process that is flexible and allows changes in user requirements during the development of the system and even after the implementation. Structured system development approaches that emphasize user participation and communication begin with a request for system study—as reflected in Table 2.2—that includes a statement of the problem as well as its requirements. During the determination of user requirements in the analysis phase, the systems analyst tries to understand the requirements and defines specifications to meet requirements. Any errors made in specifying requirements have an enormous impact on the system being developed. They inevitably lead to user dissatisfaction with the application systems and eventually the larger information system failures (e.g., In 78a).

Another key to the success of an information system that is responsive to the user is the commitment of the executive management of the user organization. It is an essential ingredient to the success of both the classical and the structured methodologies. However, the structured approach makes it easy to define the system and thus makes it easily understandable by the user management.

3.2.5 Testability of the Developed System

The purpose of testing a system is to find and correct any errors before implementation and operation. In a modular system each module is tested separately after which the integrated system is checked to ensure the proper interaction of all modules. Testing an information system before its implementation is critical because correction of any system error will be much more costly after the system is implemented and made operational. Some research results indicate that the untestability of systems requirements during the system development phase has been a major source of system errors. This is reflected in Table 3.1, which categorizes requirements and shows their relative frequency (In 78a, p. 107).

As can be seen from Table 3.1, errors due to "incorrect, untestable, or overrestrictive" requirements exceed all other errors. The point at which errors are introduced and the point at which they are discovered are especially critical in the development of computer-based information systems. This is represented in Figure 3.3 (Co 80).

TABLE 3.1   Requirements Errors in Systems Development

Type	Percentage
• Incorrect/untestable/overrestrictive	35
• Missing/incomplete	20
• Untraceable or out of scope	15
• Unclear/ambiguous	10
• Inconsistent/incompatible	10
• Typographical errors	10

The classical approach, however, does not provide the systems analyst with a means for systems testing, and it does not even consider testing as a part of system development process.

3.2.6 Other Problems

A monolithic view of the system and the "big-bang" implementation have been indicated as some of the other problems in the classical system development process (To). Because the classical systems approach takes a linear view of the application, it has a hard time dealing with large and complex systems. The big-bang occurs when an existing system is replaced by a newly developed system with no transitional period; most often the user organization will have great difficulty coping with such an event. Because of the lack of user input during the classical system development process, the user is actually introduced to the system only after the implementation so that change requests are likely to build up from that point. Thus, there is a significant gap between user expectations and system capabilities—again a main reason of user frustrations and system failures.

Another problem with the classical approach is its unsuitability to project management techniques. Poor estimations of manpower requirements may be a major reason of system problems in many applications. The estimates may be in

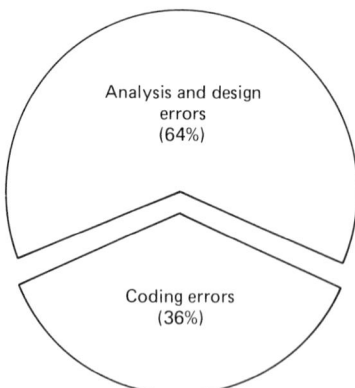

Figure 3.3 Computer-based information system errors

the form of missed deadlines, cost and resource overrun, and unsatisfactory quality of delivered systems in terms of poor maintainability, poor extensibility, poor performance in relation to cost, and low user satisfaction, for example.

Finally, inability to integrate applications into a system is another problem of classical approach. Traditionally, many applications such as payroll, accounting, and stock control, which are often tied to the existing organizations, are developed separately. Quite often serious problems of integration and dependence occur when we try to integrate such individual applications into a large, single system.

3.3 RECENT TRENDS IN INFORMATION SYSTEMS LIFE CYCLES

There have been various estimates about the cost distribution of individual steps in the classical approach to information systems development. In terms of computer-based information systems, Orr (Orr 81a) estimated that when the systems were developed using the classical approach, the planning, analysis, and design steps of the information systems life cycle took only 20 percent of the total development cost, while implementation step took almost 80 percent. Referring to other estimates given by Ramamoorthy and So (RS 78), Richardson and his colleagues (RBT 80), and Taggart (Tag 80), one can portray the hypothetical cost distribution for the classical approach as shown by the broken lines in Figure 3.4. The results obtained using classical approach have not been satisfactory, as noted before. It is indicated that "many of the problems of the software production process have arisen from a failure of those involved with software production to understand the role and nature of the design process" (In 78a, p. 8). Also "generally

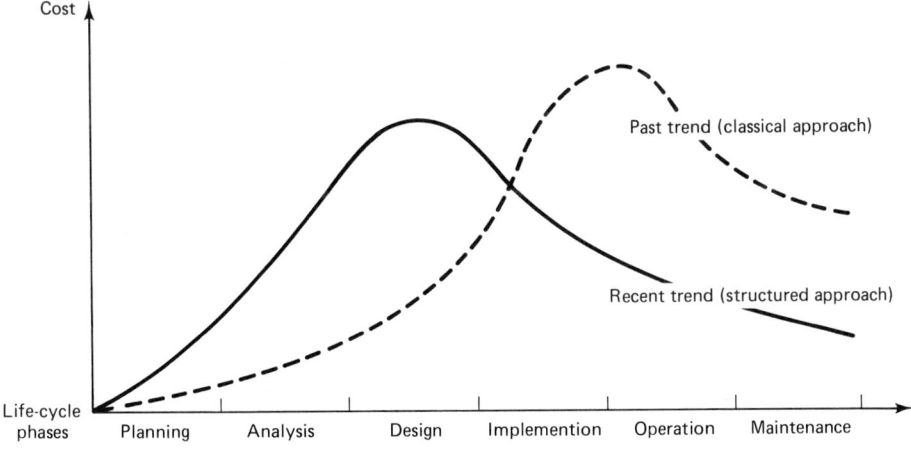

Figure 3.4   Hypothetical cost distribution of the information systems life cycle

speaking, previous experience with large scale software development has been depressing. The symptoms of the inadequacy of our software design and development methodology are high costs, inexpensive products, slippage of production schedules, and difficulties in system operation and maintenance'' (RS 78). Ramamoorthy and So (RS 78) further state that design errors account for a high percentage of the total, in fact from 36 to 74 percent. They conclude that there is a need for methods to reduce design errors. Recall that before the design phase, Table 2.2 and Figure 2.1 both descibed planning and analysis phases in the information systems life cycle. These two initial phases of systems development are sometimes termed ''requirements engineering'' (RS 78). In other research results, Ramamoorthy and So (RS 78) state that requirements problems are indeed serious and that classical means of analyzing and stating requirements are not satisfactory. The classical approach generally results in incorrect, inconsistent, and unclear requirements, causing 85 percent of the total requirements problems. Their conclusion is that a successful analysis methodology coupled with an advanced design and implementation (coding) methodology should reduce the cost elements of total development, operations, and maintenance. Taggart (Tag 80) and Orr (Orr 81a) indicate that the cost distribution pattern of the information systems life cycle has been changing. Compared to the classical approach, the earlier steps of the information system life cycle are now being allocated a higher percentage of the total development cost. This recent trend appears as the solid line in Figure 3.4.

In summary, the following tendencies can be noted:

- The analysis and design phases of information systems development processes now constitute between 70 and 80 percent of the total development cost.
- Maintenance of the existing information system claims more than half the time and effort of the technical staff of an information systems department.
- The value (in money and its effects on the total organization) of information systems has made it imperative that there must be some approaches to develop successful information systems.
- Information system development process is an engineering process, therefore similar approaches should be used.

''Structured approach'' is the name of the recent approach for information systems development. It is also called ''structured systems development methodologies'' or ''structured analysis and design methodologies'' in the literature (e.g., Ak 82).

SUMMARY

The classical systems approach is based on following the systems life cycle steps for a system development process. This has proven insufficient for developing a successful information system, which, similar to any engineering discipline, requires the proper methodologies. Such a methodology is commonly known as a "structured approach" or "a structured system development methodology."

The main problems with the classical approach for information systems development are the contractual character of the classical approach, hardware/software cost trends, high maintenance costs of existing information systems, and very limited participation of the user during development and implementation. A monolithic view of the system, "big-bang" implementation, and lack of project management capabilities have been indicated as some of the other problems in using a classical approach to information systems development.

A recent trend in information systems life cycles is that the analysis and design phases are getting more than half of the total development effort. In the past this had been just the opposite, and implementation and testing had been the major cost elements. The new trend lays special emphasis on the determination and specification of systems requirements.

EXERCISES

1. Why do we need a structured methodology of information systems development?
2. What are the major problems in using a classical systems approach for information systems development?
3. Try to obtain estimates of hardware/software cost ratios in 1960, 1970, and 1980 and indicate any trend you observe.
4. What is system maintenance? Why is it important for information systems?
5. Discuss the significance of testing in systems development.
6. Compare the recent trend in systems development to the past trend for information systems development.
7. What is the "structured approach"?

SELECTED REFERENCES

(Ak 82)   Aktas, Z. "Discussion of Structured System Analysis and Design Strategies for Information Systems." ACM Annual CS Conference, February 9–11, 1982, Indianapolis, IN.

(Bo 76)   Boehm, B. W. "Software Engineering," *IEEE Transactions on Computers*, Vol. C-25, No. 12 (December 1976), 35–50

(Bo 77)   Boehm, B. W. "Software Reliability: Measurement and Management." International Software Management Conference, London, Spring 1977.

(Ca 81)   Canning, R. G. "Easing the Software Maintenance Burden," *EDP Analyzer*, Vol. 19, No. 8 (August 1981).

(Co 80)   Connor, M. F. "Structured Analysis and Design Technique," *SoftTech, Inc.*, May 1980.

(Cr 82)   Cragon, H. G. "The Myth of the Hardware/Software Cost Ratio," *Computer*, Vol. 15, No. 12 (December 1982) 100–101.

(CW 82)   Collofell, J. S., and S. N. Woodfield. "A Project-Unified Software En-
                 gineering Course Sequence," *ACM-SIGCSE Bulletin*, Vol. 14, No. 1
                 (February 1982) 13–19.

(In 78a)   "Infotech State of the Art Report: Structured Analysis and Design," Vol.
                 1: Analysis and Bibliography, Infotech International Ltd. Maidenhead,
                 Berkshire UK, 1978.

(McGMcH 80)   McGowan, C., and R. C. Mc Henry. "Software Management," in *Re-
                 search Directions in Software Technology*, ed. P. Wegner, MIT Press,
                 1980, pp. 207–253.

(MyW 78)   Myers, W. "The Need for Software Engineering," *Computers*, Vol. 11
                 (February 1978), 12–26.

(Orr 81a)   Orr, K. "System Methodologies for the 80's," *Infosystems*, June 1981,
                 pp. 78–80.

(RBT 80)   Richardson, G. L., C. W. Butler, and J. D. Tomlinson. *A Primer on
                 Structured Program Design*, Petrocelli, 1980.

(RPTU 84)   Ramamoorthy, C. V., A. Prakash, W. Tsai, and Y. Usuda. "Software
                 Engineering: Problems and Perspectives," *IEEE Computer*, October 1984,
                 pp. 191–209.

(RS 78)   Ramamoorthy, C. V., and H. G. So. "Software Requirements and Spec-
                 ifications: Status and Perspectives," in *Tutorial: Software Methodology*,
                 ed. Ramamoorthy and Yeh, IEEE, 1978, pp. 43–164.

(Tag 80)   Taggart, W. *Information Systems*. Allyn and Bacon, 1980.

(To)   Tommela, D. R. *A Strategy for Systems Implementation*, Auerbach Pub-
                 lishers Inc., Portfolio No: 3-10-24.

Chapter 4

Structured Approach

4.1. OBJECTIVES AND MODELLING IN THE STRUCTURED APPROACH

Both problems with the classical approach and recent trends of the information systems life cycle have signalled the need for a different approach. This approach, which started to appear in early 1970s is called the "structured approach," or—more recently—the "operational approach" (Zav 84). Like the engineering approach to problem solving, the structured approach adopts some well-defined, standardized procedures and documentation, or at the least a methodology to be followed in developing a well-defined and standardized information system as the product. The resulting system will have a well-defined structure. Structure imposes order and improves comprehensibility of complex systems. It should, therefore, be an essential feature of information systems design. Structure "may pertain to the manner or form in which something is constructed or may refer to the actual system as constructed. Descriptions of structure focus on interrelation of the various parts as dominated by the general character or function of the whole. Designing structure is a process of identifying, analyzing, and selecting among alternatives with design categories" (Tau 79, p. 233).

The need for a methodology in information systems development is also expressed by Brookes et al. (BGJL 82, p. 16) as follows: "Although the systems life cycle is a useful framework within which to consider the whole systems analysis-design process, those persons responsible for carrying out the tasks need a more detailed representation or methodology to follow. Without an adequate methodology, less experienced analysts/designers may have difficulty knowing what aspect of the project should be worked on at any given time. In addition, it is usually important for all the individuals working on systems development within the one organization to follow the same procedures in terms of the sequence of steps and the means for documenting the results of their analysis and design work to assist in both project control and the interchangeability of staff.'' Structure and structured systems analysis and design concepts are not new. Assembly line techniques in factories and circuit design for electronic devices are just two examples of that idea that have been used for some time in industry. It is, however, relatively new to use these concepts in methodologies to develop information systems such that the product satisfies user needs. Through a structured approach, more complex problems of business and other organizations are solved, and the resulting system is easy to maintain, flexible, more satisfactory to users, better documented, on time, and within budget. Duran and McCready stated the major benefits of structured approaches as increased productivity, a higher quality (error-free) system, easier maintenance of the resulting systems, and greater capability of attracting and retaining quality people (DMcC 81).

Referring to Nauman et al. (NDMcK 80), one may state that to determine, define, and meet the information requirements of an organization accurately and completely is the task of the organization's information system. The most important ingredient of that system is people: managers, users, systems development personnel, and operations personnel. It is, however, difficult to obtain a correct and complete set of information requirements from an organization. Davis (DaG 82) gives three reasons for such a difficulty:

1. The limitations of human beings as information processors and problem solvers
2. The variety and complexity of information requirements
3. The complex patterns of interaction among users and analysts in defining requirements.

During system development, the information requirements of the organization are usually documented in the form of a ''functional specification'' or ''logical design'' and represent an agreement between the users and the developers of the system (NDMcK 80). This process has recently been referred to as ''requirements engineering'' in various sources and been recognized as the most important and critical part of the information systems development process.

As we shall see in the later chapters, there are various structured system development methodologies, but any of these should specify the following:

- How to recognize a good design
- How to create a good design
- How to communicate a good design

In addition, an effective methodology will incorporate purposeful structuring and modularity of the system under consideration. The resultant information system will then have most, if not all, of the following properties (e.g., DMcC 81, RDMcG 77, Sch82, TPW 81):

- Acceptability—users find the quality and efficiency satisfactory
- documentation—clarity of goals and methods documented during development results in better communications among users, developers, and managers
- testability—chances of future failures and/or user dissatisfaction are minimized
- cohesiveness—maximum interaction within each component (module)
- compatibility—the system "fits" the total, integrated system
- economy—the system should be cost effective within the given resources
- efficiency—resource utilization is optimal
- fast development rate—relatively less time is needed for development
- feasibility—resultant system should satisfy all feasibility criteria
- flexibility—it is easy to modify, add, or delete components
- logic/hierarchy—components of the system are logically/hierarchically related to each other
- low degree of coupling—there is minimum interaction between components (modules)
- maintainability—future maintenance and enhancement times and efforts are reduced
- modularity—the system has relatively independent and single-function parts that can be put together to make a complete system; modularization, decomposition, parsing are the terms that are used interchangeably
- reliability—error rate is minimized; outputs are consistent and correct;
- visibility—it is easy to perceive how and why actions occur (they are traceable)
- simplicity—complexity and ambiguity are minimized
- timeliness—the system should operate well under normal, peak, and recovery conditions
- uniformity—the structure of the components (modules) should be uniform
- user friendliness—the system meets user needs and acts as a catalyst in achieving objectives.

The above properties may be stated as the objectives of a structured systems development methodology for an information system; they are summarized as Table 4.1.

Structuring the information system during development—that is, during the analysis and design phases, as well as during implementation, operation, and

TABLE 4.1   Some Properties of a Structured
Information System

• acceptable	• Logical/hierarchical
• better documented	• low coupling
• better tested	• maintainable
• cohesive	• modular
• compatible	• more reliable
• economical	• observable/visible
• efficient	• simple
• fast development rate	• timely
• feasible	• uniform
• flexible	• user friendly

maintenance—has been the common property of a group of methodologies known as "structured methodologies."

Although it is now common to use structured documentation, structured testing, structured walk throughs, structured validation and verification, and structured project management during an information system development process, the crucial phases are structured analysis (or requirements specification) and structured design.

As stated by Duran and McCready (DMcC 81), structured analysis is used to define and describe the system that best satisfies user requirements, given certain time and budget constraints. The objective of structured design is to minimize the lifetime cost of an information system by emphasizing maintainability, because maintenance is the most costly component in a system's life cycle. The critical point in design is to match the solution to the problem.

Although there are a variety of ways to implement structured analysis and structured design, some of the features all methodologies share are the use of graphical models, emphasis on user communication (and hence user involvement), repetition of the previous phase(s) and step(s), and reviews. In the structured analysis process, the models should represent the functions of the system rather than the means to accomplish them; in other words, emphasis is on the logical components of the system rather than its physical components. Consideration of design and implementation issues is postponed until agreement has been reached between designers and users on the function, or objectives, of the system. In some structured design methodologies, a set of evaluative criteria is also provided as part of the methodology as a kind of checklist for the systems analyst.

The models that are used in structured analysis and design and their interrelation are depicted in Figure 4.1. The first model describes how the current system operates by depicting its physical components. The second model is a logical abstraction of the current system, prepared to show what the existing system does. The third model represents the planned operation of the new system. By adding general physical components to the logical model of the new system we then arrive at a physical model of the new system—number 4 in Figure 4.1—that details the actual implementation.

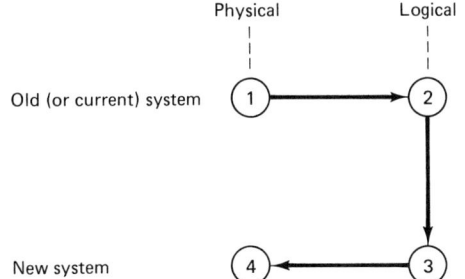

Figure 4.1 Models in structured analysis and design

4.2 SOME TERMINOLOGY

A number of terms are commonly used in the structured approach and need to be clarified at this point. The dictionary (Web 75) defines these terms as

algorithm	A step-by-step procedure for solving a problem
method	A systematic way, technique, or process of or for doing something; a body of skills and techniques
methodology	A body of methods, rules, and postulates employed by a discipline
strategy	The art of devising or employing plans toward a goal
technique	The manner in which technical details are treated
tool	Something (as an instrument or apparatus) used in performing an operation or necessary in the practice of a vocation and profession.

Freeman (Fr 79) defined a method as a way of doing something; by extension, methodologies are collections of methods and tools chosen to complement one another, along with the management and human factor procedures necessary for their application. Techniques are defined as informal methods and tools as objects such as programs, languages, or documentation forms that help us use a method. Davis (DaG 82) defined a method as an orderly or systematic procedure and a methodology as a set of methods and techniques. He also stated that strategies are general approaches for achieving objectives and that methods and methodologies are the detailed means for doing it. Comparing the terms ''method'' and ''methodology,'' Infotech Report (In 78a) notes that a method is a procedure for carrying out a particular task and that a methodology consists of an integrated set of methods, based on a reasoned set of basic principles, along with rules for applying them. Griffiths (Gr 78) observes that the distinction between method and methodology has never been made explicit and indicates that a design methodology provides, in some measure, reasons for all the steps in the design process; furthermore, these reasons may be understood completely without reference to a particular application.

As is clear from the definitions on p. 33, the terms "method," "methodology," and "strategy" have very close meanings. For our purposes in this text, we shall use the term "method" to signify a way to solve a problem; strategy and methodology will be taken as synonyms to mean a group of methods and tools along with certain rules for the development and operations of an information system. Also, the phrase "structured system development methodology" will be used to mean "structured approach" and vice versa.

4.3 STRUCTURED TOOLS

The tools that are used in the structured approach are sometimes grouped as analysis and design tools. Howden (How 82) also mentions some verification tools—such as file comparator, test harness, and performance monitor–and management tools and techniques such as automated project control system, project control system, and project status report generators.

A shared property of the tools that are used in the structured approach is that they are graphical. There are also some nongraphical tools. In fact, the phrase "a picture is worth a thousand words" is the basic idea behind most of the tools and methodologies that are employed in the structured approach.

Another common characteristic of most of the structured tools is that they are based on the "tree concept." One may recall the applications of trees for data and expression representation or as decision and game trees. If one examines the structured tools of such methodologies as the hierarchy diagram, structure chart, Jackson diagram, or Warnier/Orr diagram, it will become evident that they are just an application of the tree concept.

Many of the tools that will be discussed in the following chapters are components of their respective systems development methodologies. SADT, HIPO, the Jackson diagram, Warnier/Orr diagram, data flow diagram (DFD), and the structure chart are some of the examples of that category.

The literature reveals several attempts to classify the tools of the structured approach. For example, Peters (Pes 81) classifies them as architectural, structural, behavioral, and database techniques. The difficulty with such a classification is that often a particular tool will easily fit several categories. We, therefore, prefer to classify them broadly as graphical and nongraphical tools and to treat HIPO, data flow diagrams, structure charts, SADT, Jackson Diagrams, the entity-relationship model, and Warnier/Orr Diagrams as the typical graphical tools. Data Dictionary, Structured English, and pseudocode will be treated as typical nongraphical tools that are used in structured approach.

Almost all of the tools that are used in the structured approach are identical to those that are proposed and used for software development. A thorough survey and classification of such tools is given in a U.S. Department of Commerce publication (Hou 82).

Another tool that is often used together with structured tools is the classical flowchart. Flowcharts are used to describe the programming logic of a problem but also to describe the physical components of an information system. Such a flowchart is called a "system flowchart," and some of the symbols that are used in it are tabulated in Figure 4.2.

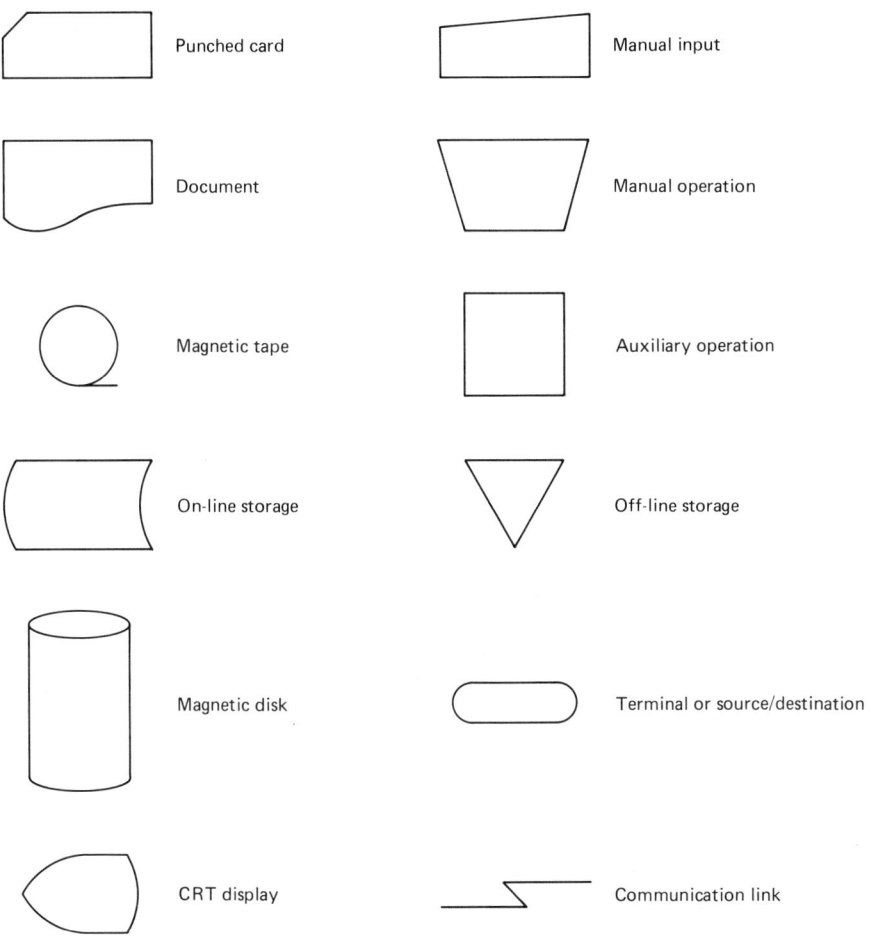

Figure 4.2 System flowchart symbols

4.4 STRUCTURED METHODOLOGIES

There are numerous methodologies available for the development of information systems. A partial listing of available packaged information systems development methodologies is given by Benenati (Be 82). Atena, Auto Flow, Cara Systems

Development Standards, Dauphin, Domonic, Glossary, HIPO, Jackson Design, Prosim (Long Range Planning Models), Mentor Systems Project Methodology Plus, Nichols Project Management Control System, Odyssey, PACII, PRIDE, Program Design Management (PDM), Project Control/70, (System Development Methodology) SDM/70, PROVOC, PSL/PSA, PSDM, SADT, Software Life Cycle Management (SLCM), Spectrum-1, SREM, Structured Design, and Warnier/Orr Diagrams are some of these packages.

Unfortunately, the industry has not yet established clear standards. However, some methodologies are being successfully used in a number of organizations, among them Structured Design, Warnier/Orr, Jackson System Development, and SADT. In the third part of the book, the most commonly used methodologies will be grouped together and the typical ones of each group will be discussed. Applications of some of these methodologies will further be demonstrated on a case study presented in the Appendix.

4.5 OTHER ISSUES PERTAINING TO THE STRUCTURED APPROACH

Another point for discussion has to do with the type of user, or the potential application, of the structured approach. From time to time certain methodologies are proposed exclusively either for business systems or for nonbusiness areas such as engineering. It is our belief that a good system development methodology should be applicable to any system, business or nonbusiness.

A few words are now in order about the psychological dimension of the information system development process. Remember, it has already been stated that people are the most important ingredient of an information system. It is, therefore, no surprise to witness a resurgence of interest in the psychological dimension of information systems development (e.g., Mor 81). Another human factor is Miller's magical number seven (Mil 56). In that paper, he was referring to the seven wonders of the world, the seven primary colors, the seven basic notes of the musical scale, seven days of the week, the seven categories for absolute judgment, the seven objects in the span of attention, and the seven digits in the span of immediate memory. The number seven has a special meaning in some religions too. For example, seven is a ''perfect number'' in Islam and in the Judaeo-Christian tradition links the threefold nature of the cosmos with the four earthly elements; water, air, fire, and earth. In structured systems design methodologies 7 ± 2 has been the number that is used for limiting the number of modules and/or the number of hierarchical levels of the system being developed in order to deal with the issue of complexity and improve the clarity of the resulting information system. The subject of system complexity and its relation to human capacity is another recent important issue relevant to structured systems development methodologies for information systems (Ze 82).

Next we would like to address a criticism about the structured system development methodologies for information systems. Recently Kimmerly (Ki 82) claims that "our inability to match hardware advances lies, at least partially, in the failure of both practicing systems analysts and computer science academicians to stress adequately the importance of aesthetics, imagery, and other precursors of creativity in the methodology of the discipline, particularly with respect to definition of problems and the conceptualization of solutions. As a result, a significant imbalance now exists between the emphasis being placed on high levels of structure on the one hand and creativity on the other." He later concludes that "due in part to the legacy of the various 'structured revolutions' creativity has not only been comprehensively de-emphasized, but has come to be regarded as something to be avoided altogether."

It is appropriate at this point to recall the "art vs. science" debate in the area of programming languages in the late sixties and early seventies. As we all know the answer to that debate has been structured programming, a methodology that is both teachable and learnable. Such a program is much more efficient and valuable to an organization than "clever or tricky or artistic" programs. Similarly, structured systems development methodologies for information systems are proving that the information systems development process can be teachable and learnable, just the same as any engineering system development process. Of course, however, there is always a need for human judgment at critical decision points. Just as "engineering judgment" is an integral part of developing an engineering system, so there is always a need for systems analysts to interject their opinions and judgments during the information systems development process. This is where a certain degree of creativity is essential in structured information systems development.

SUMMARY

Sometimes an approach for information system development is labelled "structured" if the classical information system life cycle-steps are followed closely and some relevant tools (known as structured tools) are used during these steps. Mostly, however, a structured methodology refers to a strategy that will yield a successful information system. Such a stragety applies to the analysis and/or design phase of systems development and employs structured tools and a life cycle concept as in the classical approach.

The major objective of a structured systems development methodology for an information system is to produce a well-defined and standardized information system using well-defined and standardized procedures and documentation. The resulting information system has most of the following properties: It is acceptable, better documented, better tested, cohesive, compatible, economical, efficient, feasible, flexible, hierarchical, maintainable, modular, more reliable, observable, simple, timely, uniform, user friendly, and has a fast development rate and low coupling. System modelling is very important in the structured approach, which employs both logical and physical models.

Algorithm, method, methodology, strategy, technique, and tool are some terms that are used relevant to the structured development processes.

Structure is an essential feature of a development process, since structure imposes order and hence improves the comprehensibility of complex systems. Structuring of the information systems during development, that is, during the analysis and design phases, as well as during implementation, operation, and maintenance, has been the common property of a group of methodologies commonly known as "structured methodologies." There are numerous structured analysis and design methodologies. Use of graphical models, emphasis on user communication and hence user involvement, nonlinearity or iteration, and repeated reviews are some of their common features.

Almost all of the tools that are used in the structured approach are those that are proposed and used for software development. In addition, system flowcharts may be used together with structured tools.

There are a number of structured methodologies for information systems developments. Some of them are manual and some are computerized.

The psychological dimension of the information systems development process is getting more attention lately. Some of the criticism levelled at the structured approach reminds us of the "art vs. science" debate in the area of programming languages in the late sixties. The most important feature of any structured systems development methodology is that it is teachable and learnable.

EXERCISES

1. What is meant by the "structured approach"?
2. What is the main objective of a structured system development methodology?
3. What are the properties of a structured system? Explain them.
4. What is the significance of structure to a development process?
5. What are the common features of structured analysis?
6. What are the common features of structured design?
7. What is the relationship between physical and logical models in structured analysis and design?
8. Define the terms *algorithm*, *method*, *methodology*, *strategy*, *technique*, and *tool*.
9. What is the difference between a method and a methodology?
10. What is the difference between a methodology and a strategy?
11. What are some properties of structured tools?
12. Give any two examples of graphical and nongraphical tools.
13. What is the main reason for using a system flowchart?
14. Give any two examples of structured methodologies.

15. Is there a need for different methodologies for business versus nonbusiness applications? Discuss.
16. What is the idea behind the magical number 7?
17. What is the significance of the human factor in information systems development?
18. What is "requirements engineering"?

SELECTED REFERENCES

(BGJL 82)   Brookes, C. H. P., P. J. Grouse, D. R. Jeffery, and M. J. Lawrence. *Information Systems Design*, Prentice-Hall, 1982.

(DaG 82)   Davis, G. B. "Strategies for Information Requirements Determination," *IBM Systems Journal*, Vol. 21, No. 1 (1982), 4–30.

(DMcC 81)   Duran, P., and A. McCready. *Structured Techniques*. Auerbach Publishers, Inc., Portfolio No: 3-10-20, 1981.

(Fr 79)   Freeman, P. *A Perspective on Requirements Analysis and Specification*. Auerbach Publishers, Inc., Portfolio No: 32-04-01, 1979.

(Gr 78)   Griffiths, S. N. "Design Methodologies," in *Info Tech State of the Art Report: Structured Analysis and Design*, Vol. 2, p. 139, Infotech International Ltd., Maidenhead, Berkshire, UK 1978.

(Hou 82)   Houghton, R. C. *Software Development Tools*, NBS Special Publication, No: 500-88. U.S. Dept. of Commerce, March 1982.

(How 82)   Howden, W. E. "Contemporary Software Development Environments," *Comm. ACM*, Vol. 25, No. 5 (May 1982), 318–29.

(In 78a)   *Info Tech State of the Art Report: Structured Analysis and Design*, Vol. 1: Analysis and Bibliography, Infotech International Ltd., Maidenhead, Berkshire, UK 1978.

(Ki 82)   Kimmerly, W. C. "Restricted Vision," *Datamation*, November 1982, 152–60.

(Mil 56)   Miller, G. A. "The Magical Number Seven, Plus or Minus Two: Some Limits on Our Capacity for Processing Information," *Psychological Review*, Vol. 63, No. 2 (March 1956), 81–97.

(Mor 81)   Moran, T. P. "An Applied Psychology of the User," *Computing Surveys*, Vol. 13, No. 1 (March 1981), 1–11.

(NDMcK 80)   Naumann, J. D., G. B. Davis, and J. D. McKeen. "Determining Information Requirements," *The Journal of Systems and Software*, No. 1 (1980), 273–81.

(Pes 81)   Peters, L. J. *Software Design*. Yourdon Press, 1981.

(RDMcG 77)   Ross, D. T., M. E. Dickover, and C. McGowan. *Software Design Using SADT*, Auerbach Publishers, Inc., Portfolio No: 35-05-03, 1977.

(RPTU 84)   Ramamoorthy, C. V., A. Prakash, W. Tsai, and Y. Usuda. "Software Engineering: Problems and Perspectives," *IEEE Computer*, October 1984, 191–209.

(Sch 82)   Schach, S. R. "A Unified Theory for Software Production," *Software Practice and Experience*, Vol. 12 (1982), 683–89.

(Tau 79)   Tausworthe, R. C. *Standardized Development of Computer Software*, Part II, *Standards*. Prentice-Hall, 1979.

(TPW 81)   Thayer, R. H., A. B. Pyster, and R. C. Wood. "Major Issues in Software Engineering Project Management," *IEEE Transactions on Software Engineering*, Vol. SE-7, No. 4 (July 1981), 333–42.

(Web 75)   Webster's New Collegiate Dictionary, G. & C. Merriam, 1975.

(Zav 84)   Zave, P. "The Operational Versus the Conventional Approach to Software Development," *Comm. ACM*, Vol. 27, No. 2 (February 1984), 104–18.

(Ze 82)   Zeigler, B. P. "Systems Complexity, System Methodology and the Human Capacity to Manage Them," Wayne State University, Department of Computer Science, 1982.

Chapter 5

Hierarchy Charts and HIPO

5.1 INTRODUCTION

A hierarchy chart is also called a function chart. As its name implies, a hierarchy chart shows the hierarchical relations of the modules (i.e., components) of a system under consideration. In such a diagram, each individual module of the system is described by its main function in terms of a verb and an object. As an example, consider the basic modules of a program that updates an inventory master file in Figure 5.1. As seen in Figure 5.1, a hierarchy chart is quite similar to an organization chart where each diagram at any level is a subset of the level above it. A hierarchy chart shows the modular hierarchy of a system as well as its modular partitioning and functions. For example, data retrieval, process, and write functions and their subfunctions are shown in Figure 5.1. However, major loops and decisions involved in a system and communication within the modular hierarchy of the system are not shown.

Well-structured programs or systems are most commonly characterized as "hierarchical" or "tree-structured." As noted by Turner (Tu 80), the distinction between hierarchy and tree is rarely made. He defines a hierarchy as any program (or system) structured in levels, where the modules on a given lower level may or may not be shared by the modules on the higher levels. Tree is then defined as a specific type of hierarchy where sharing is not allowed.

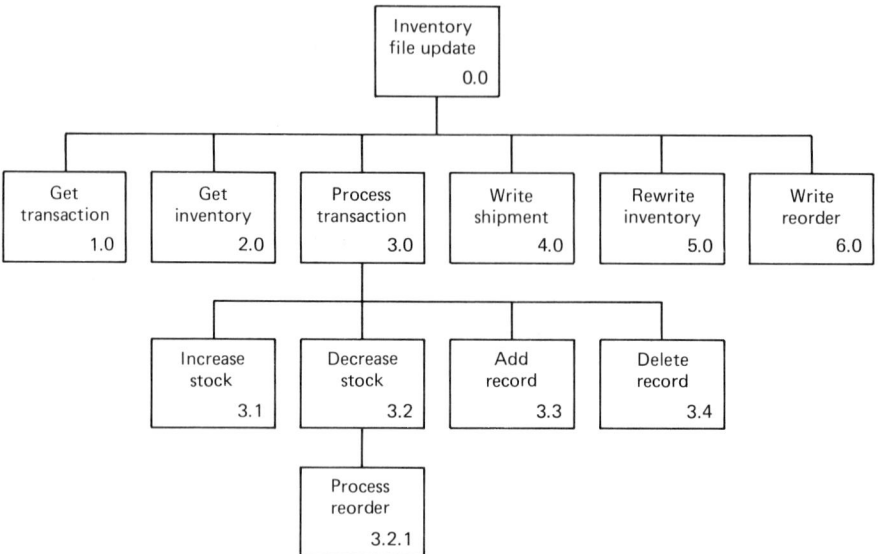

Figure 5.1   Basic functions to update an inventory file

5.2 HIPO

HIPO is an acronym for Hierarchy plus Input-Process-Output. It is a package that consists of a set of diagrams that graphically describe the functions of a system from the general level to the detail level. Initially each major function is identified and then subdivided into lower-level functions. HIPO is developed and supported by IBM (Au, IBM 75, Ka 76).

5.2.1 Uses and Objectives of HIPO

HIPO, similar to many other structured tools of information systems development, is originally a program documentation tool. Flowcharts used in programming describe the program logic. The functions of the program modules, that is, what they do, is not depicted in a flowchart. As noted in an IBM publication (IBM 75), some IBM personnel believe that programming systems documentation based on function can contribute to the efficiency of a program maintenance effort by speeding the location in the code of a function to be modified. HIPO is thus developed as a technique to document functions of programs. HIPO is now also used as a design aid and documentation technique throughout the information system life cycle.

The major objectives of HIPO as a system development and documentation technique may be summarized as (IBM 75):

1. to provide a structure by which the function(s) of a system can be understood
2. to state the functions to be accomplished by a program rather than specify the program statements to be used to perform the functions
3. to provide a visual description of input to be used and output produced by each function for each level of diagrams

The most important objective of a system is to produce output that is correct and meets users' requirements. A HIPO diagram allows one to see how input to a system is transferred into output. Automated programs such as HIPODRAW are available to provide computer-generated and computer-maintained HIPO documentation. Such a program accepts user-written statements describing the desired HIPO documents and produces on a standard line printer all components of a HIPO package.

5.2.2 Kinds of Diagrams in a HIPO package

A typical HIPO package contains three kinds of diagrams:

1. Visual Table of Contents (VTOC): One or more hierarchy diagrams.
2. Overview diagrams: A series of functional diagrams, each of which is related to one function of the system.
3. Detail diagrams: A series of functional diagrams each of which is related to one subfunction of the system.

These diagrams may further be summarized as follows (IBM 75):

1. Visual Table of Contents (VTOC) or hierarchy diagrams:
   Such a diagram contains the names and identification numbers of all the overview and detail HIPO diagrams in the package and shows the structure of the diagram package and relationship of the functions in an hierarchical fashion. A description section is also included to describe each function. (See Figure 5.2 as an example.)
2. Overview Diagrams
   These are high-level HIPO diagrams that describe the major functions and reference the detail diagrams needed to expand the functions to sufficient detail. Overview diagrams provide, in general terms, the inputs, processes, and outputs of a particular function. The input section contains those data items that are used in the process section. The process section contains a series of numbered steps that describe the function being performed. Arrows connect the input data items to the process steps. The output section contains those data items that are created or modified by the process steps. Arrows

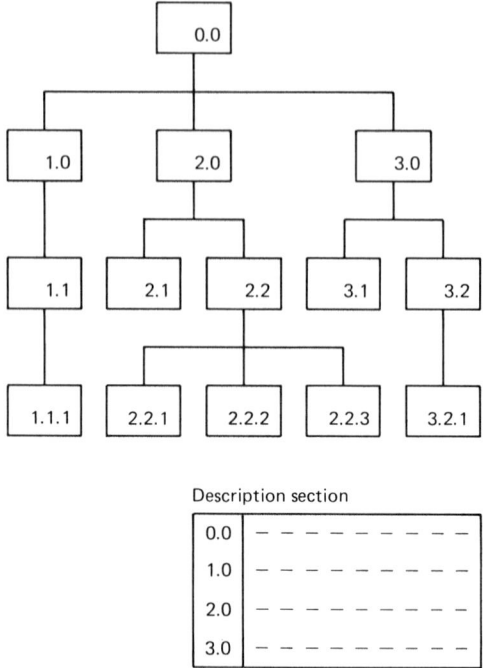

Figure 5.2 Visual table of contents (or hierarchy diagram)

connect the process steps to the output data items (see Figure 5.3a). An extended description area is included in an overview diagram and can amplify the process steps and input and output data items. The extended description also refers to lower level HIPO diagrams, non-HIPO documentation, and code. As is stated in an Auerbach Portfolio (Au), extended description is a table containing prose explanations of a process depicted in the diagram. The extended description can be on the same page as its associated diagram or on an adjacent sheet, depending on the available space. Entries in the extended description are keyed to the detail diagrams by a numbering scheme. Entries also contain cross-references to other documentation (flowcharts, program specs, and so on) or to the implementation itself (module names, subroutines, or labels). The extended description feature of the HIPO package thus serves to tie together the diverse elements of system documentation that may be produced in addition to the diagrams.

3. Detail Diagrams

These are lower level HIPO diagrams that contain the fundamental elements of the package. They describe the specific functions, show specific input and output items, and refer to other detail diagrams. Similar to overview diagrams, detail diagrams may also have an extended description (see Figure 5.3b).

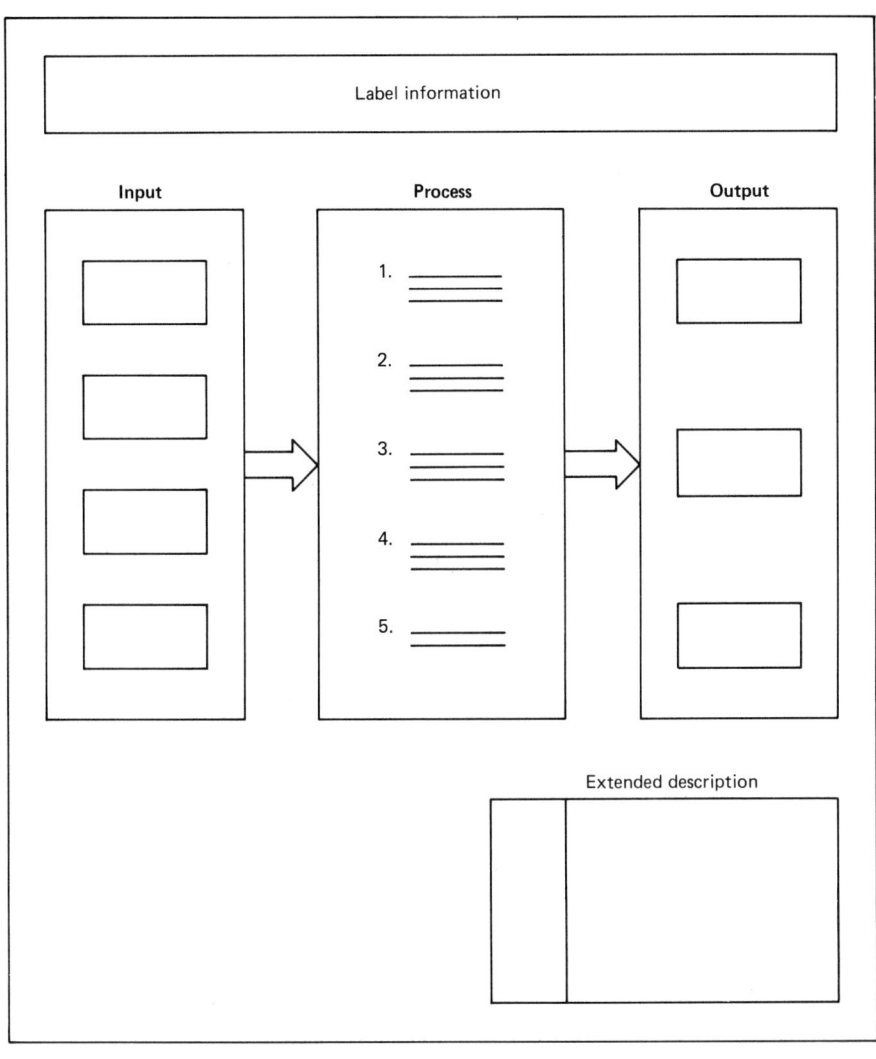

(a)

Figure 5.3   (a) Overview diagram, (b) Detail diagram

No hard-and-fast rule exists to determine the maximum number of levels permitted in the hierarchy diagram (i.e., VTOC); however, the 7 ± 2 rule appears to set a practical limit beyond which the package becomes cumbersome. The number of levels is generally a function of the complexity of the system being documented.

Two available aids for preparing HIPO diagrams are the HIPO Worksheet (Figure 5.4) and the HIPO Template (Figure 5.5), both provided by IBM.

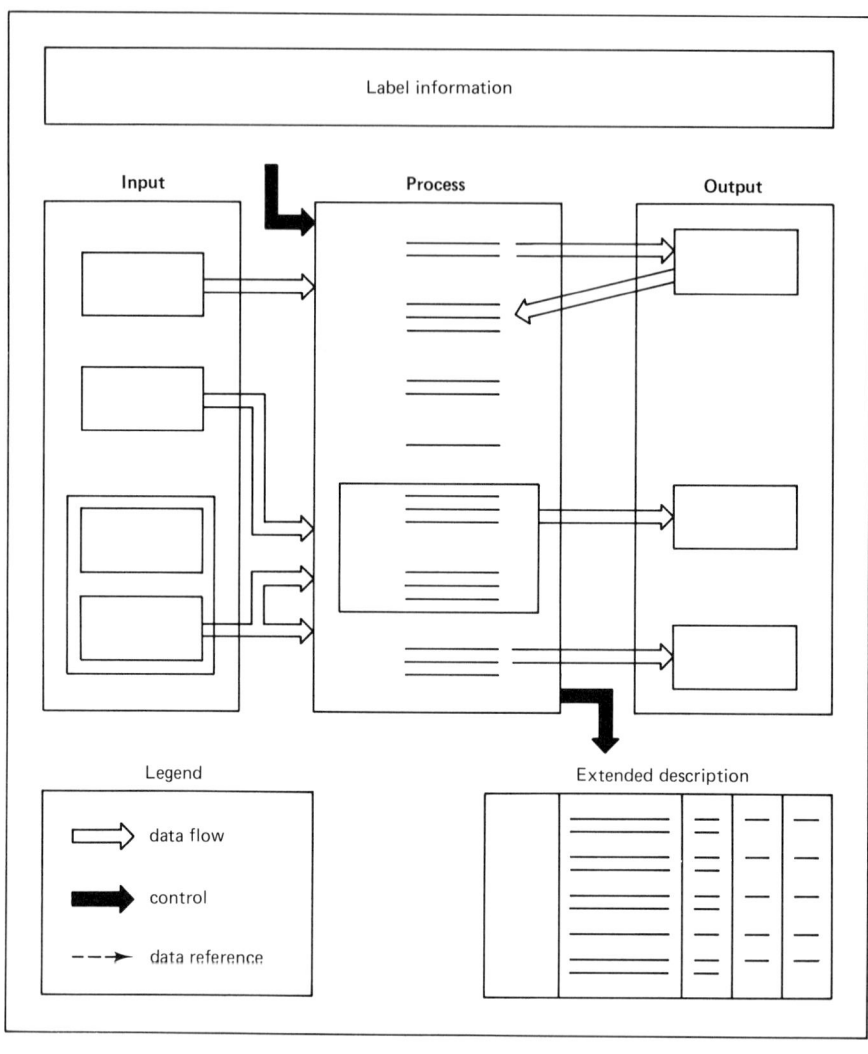

(b)

Figure 5.3   *(cont.)*

The strongest features of HIPO are

- its simplicity
- the ease with which users can learn it
- the efficiency of analyst-user communication

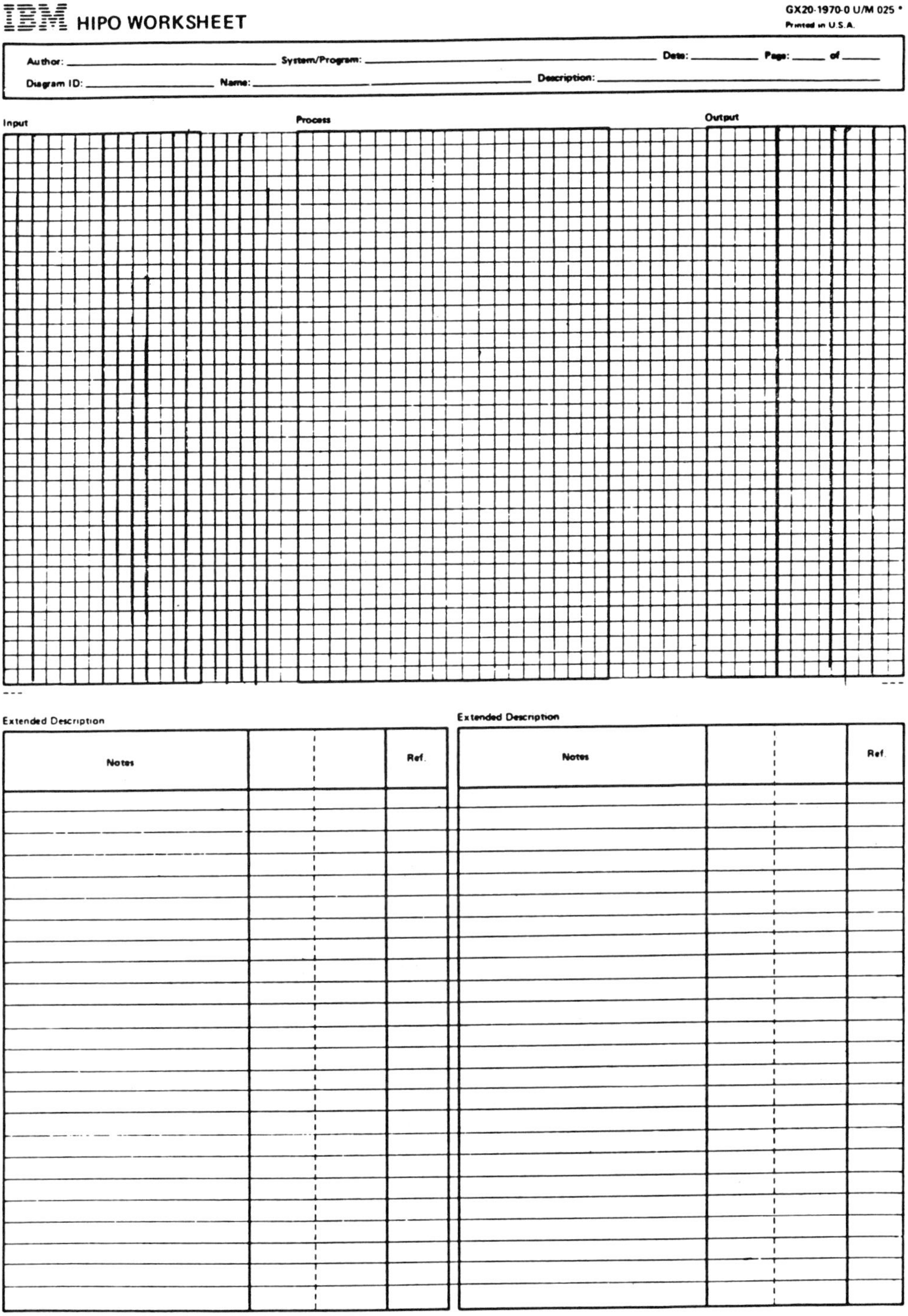

IBM HIPO WORKSHEET
GX20-1970-0 U/M 025 *
Printed in U.S.A.
Author:
System/Program:
Date:
Page:
of
Diagram ID:
Name:
Description:
Input
Process
Output
Extended Description
Notes
Ref.
Extended Description
Notes
Ref.

Figure 5.4 HIPO worksheet

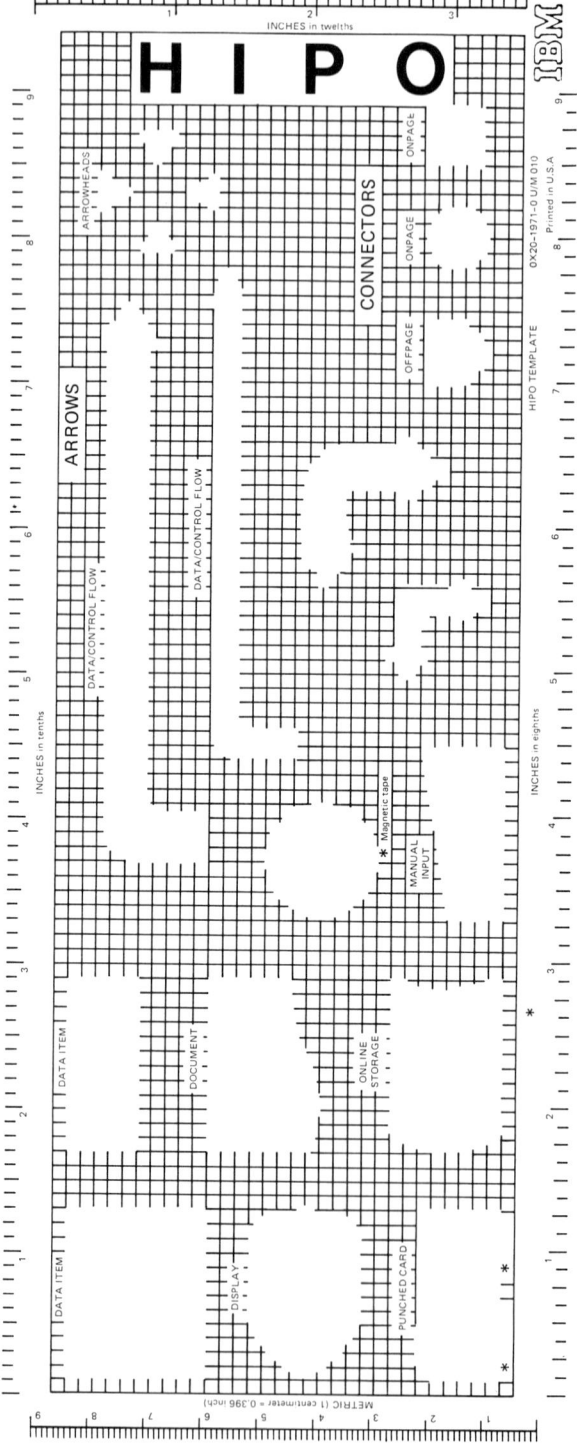

Figure 5.5 HIPO template

48

There are also some challenges in using HIPO. Although users can comprehend the terms used in the notation, large systems of HIPO diagrams are very difficult to change manually and to verify for consistency. Also, data feedback is difficult to describe.

HIPO has been used as a program documentation tool in a variety of business applications. However, the rippling effect (changes in related diagrams associated with changes to systems) can make HIPO impractical for very large system applications. This becomes particularly true when HIPO is used as a system architecture documentation tool.

5.2.3 An Example

The following example illustrates the three different kinds of diagrams in a HIPO package: the hierarchy diagram, overview diagram, and a detail diagram (see Figures 5.6–5.8). The problem relates to a traffic control system.

5.2.4 Steps in Creating HIPO Diagrams

The following steps may be suggested as a guide in drawing both overview and detail diagrams:

1. Prepare empty input, process, and output boxes of HIPO sheets.
2. List any known output in the output box.
3. Develop the contents of the input and process boxes next and fill in any intermediate output not previously specified.
4. Using as few words as possible, state each function in the process box.
5. Connect the corresponding input items and process steps with arrows; connect the corresponding process steps and output items with arrows.
6. Try to combine related data items into logical groupings using boxes.

Some visually important points in developing HIPO diagrams are (IBM 75):

- Arrows generally go from left to right adjacent to the referenced or connected items, and from items higher on the page to those lower on the page.
- Where the tail of the arrow starts and where the tip of the arrow ends must be clear.
- Arrow types must be consistent with the legend.
- Boxes should contain only functionally related items.
- Abbreviations should be avoided in overview diagrams.

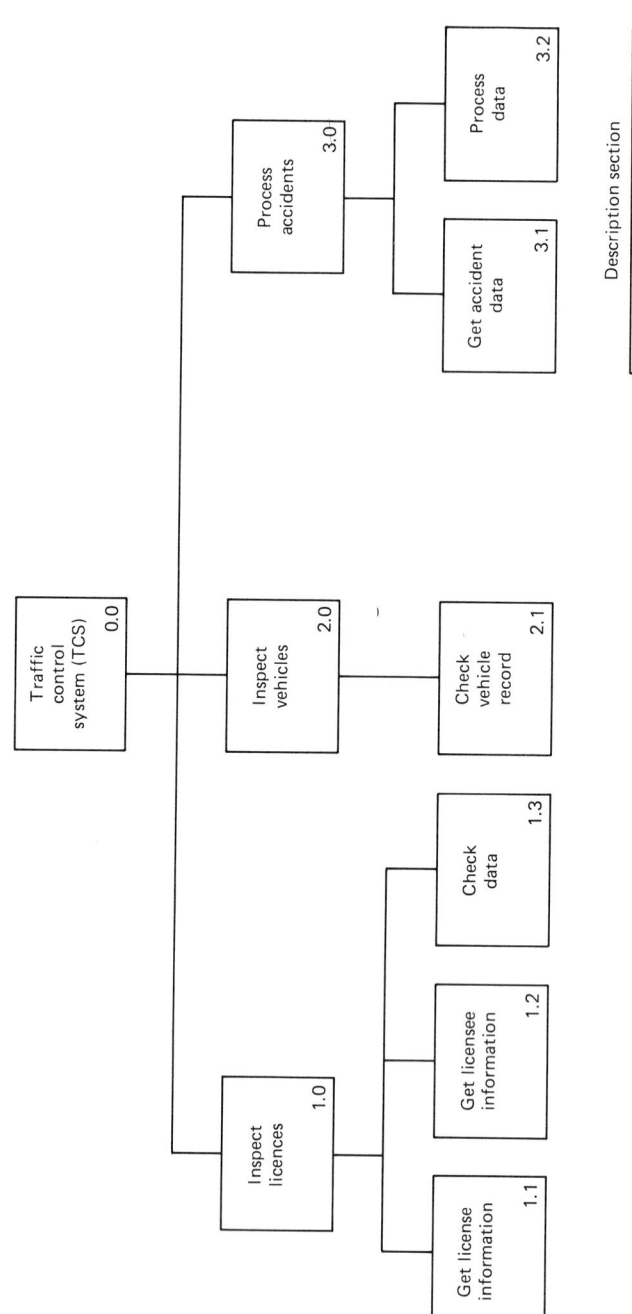

Figure 5.6 Hierarchy diagram example

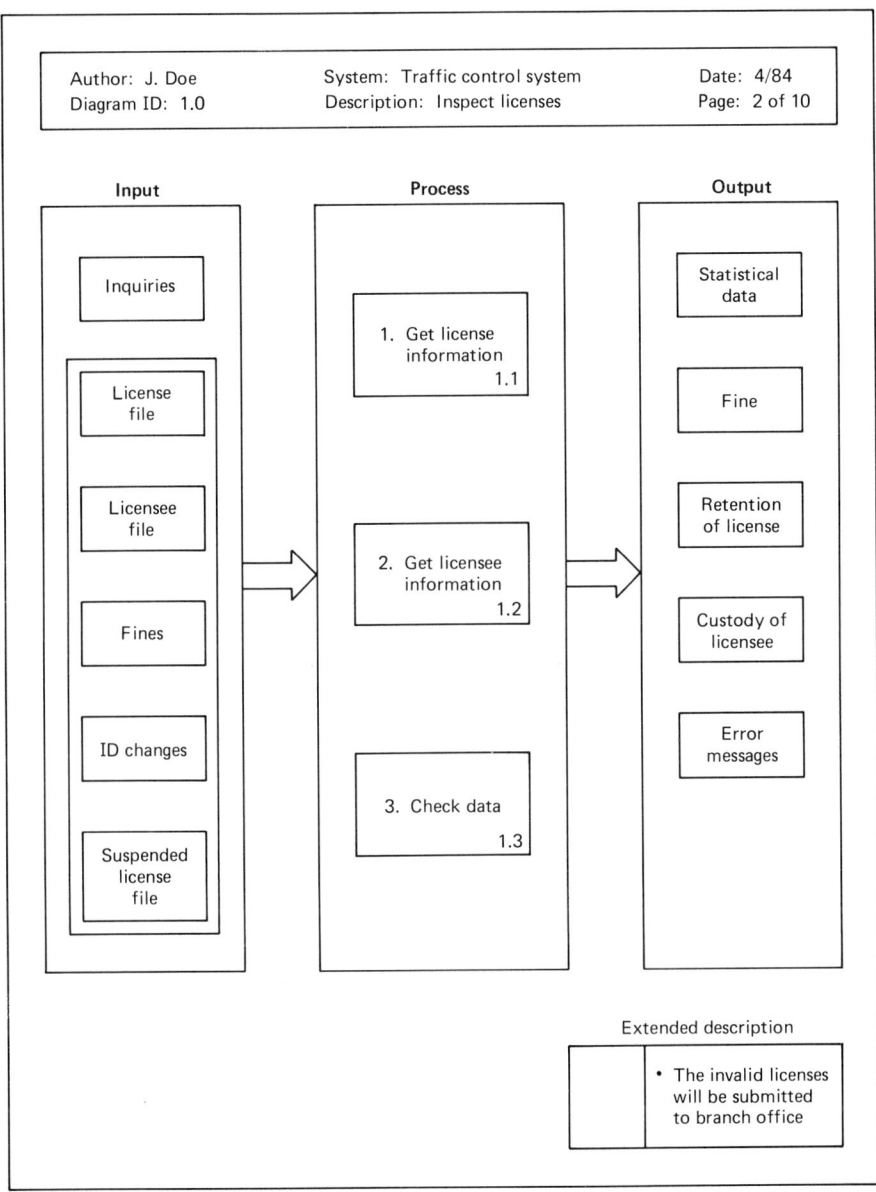

Figure 5.7   Overview diagram example

- All terms and labels should be defined in the extended description when first used.
- Data in overview diagrams should be generally described to minimize arrows.

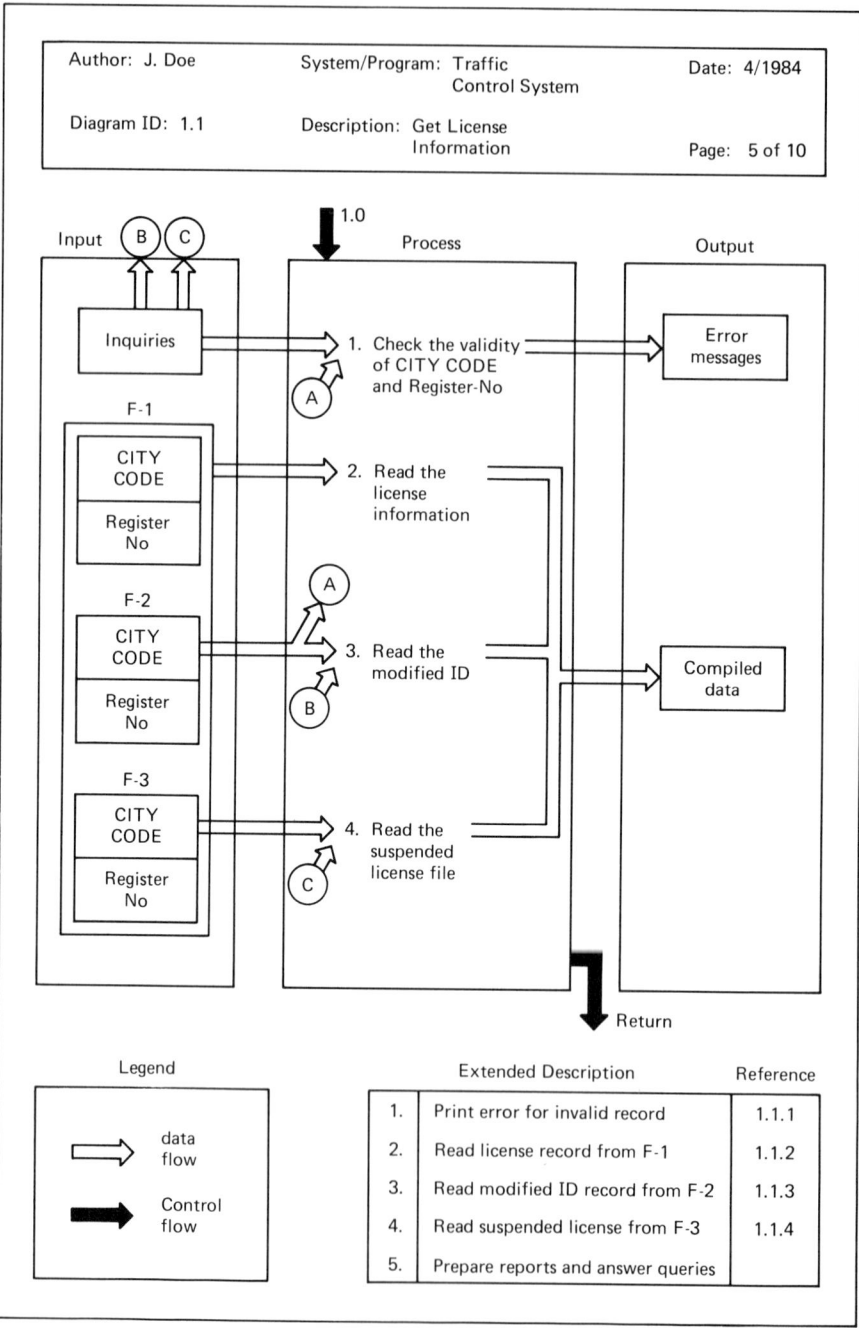

Figure 5.8   Detail diagram example

5.2.5 Types of HIPO Packages

HIPO diagrams are a means for system documentation. During systems development the output from one phase is input to the next phase of the systems life cycle.

Basically, there are two kinds of HIPO packages: the initial design and the detail design. An optional third kind, the maintenance package, is sometimes used. See Figure 5.9.

Initial design package: The designers communicate and validate their ideas by using HIPO diagrams. This package is then used for design reviews by management and other groups, including users.

A review of the initial design package should look for content, consistency, and both technical and functional accuracy.

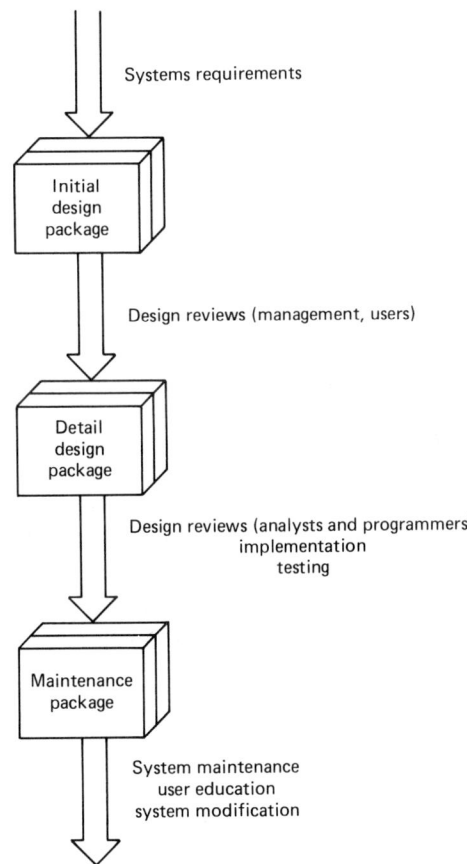

Figure 5.9   Three kinds of HIPO packages

Detail design package.   Using the initial design package as a base, analysts and programmers now add the details, including more levels to the HIPO diagrams, and use the resultant package for implementation and comparison with the initial design package to ensure that all requirements have been met. Frequently this package is the final HIPO documentation and is used as the maintenance package.

In addition to the points that are looked for in the review of the initial design package, appropriate names should be given to such items as programs, files, and tables that are used consistently in the package. Detail design package diagrams should be compared to initial design diagrams to avoid any error.

Maintenance package.   This package is used for corrections, changes, or additions to the system.

As is shown in Figure 5.9, the detail design package is input to the maintenance phase of system development. Reviews of the maintenance phase HIPO package may be conducted for the following purposes:

1. To educate user personnel
2. To check the HIPO package for clarity, legibility, and completeness
3. To optionally delete or add some low-level diagrams or to make some modifications

5.2.6 HIPO and Its Relations to Some Other Tools

Some structured tools such as top-down development, structured programming, chief programmer team, and structured walk-throughs are usually referred to as IPT (Improved Programming Technologies) of IBM.

Combining top-down development with structured programming results in a software of extreme modularity, and HIPO diagrams are valuable and practical tools for such a development effort. Again, the use of HIPO diagrams is reported to be beneficial for structured walk-throughs and for chief programmer team organizations, both of which are used with top-down development and structured programming. Combining top-down development and structured programming results in a program of extreme modularity both in function and logical structure. HIPO diagrams are a logical extension of the functions identified in top-down development and provide the necessary documentation from the start of a project through implementation. Another concept that is being used with top-down development and structured programming is the chief programmer team organization. Structured walk-throughs have been implemented within programming groups that rely on top-down development, structured programming, and chief programmer teams.

5.3 A VARIATION OF HIPO DIAGRAMS

Overview diagrams and detail diagrams of a HIPO package are usually referred to as IPO Diagrams. Instead of using a different format for these diagrams from those shown in Figure 5.3, some authors have reported using the same format for both. Figure 5.10 represents such an alternative format (DaW 83).

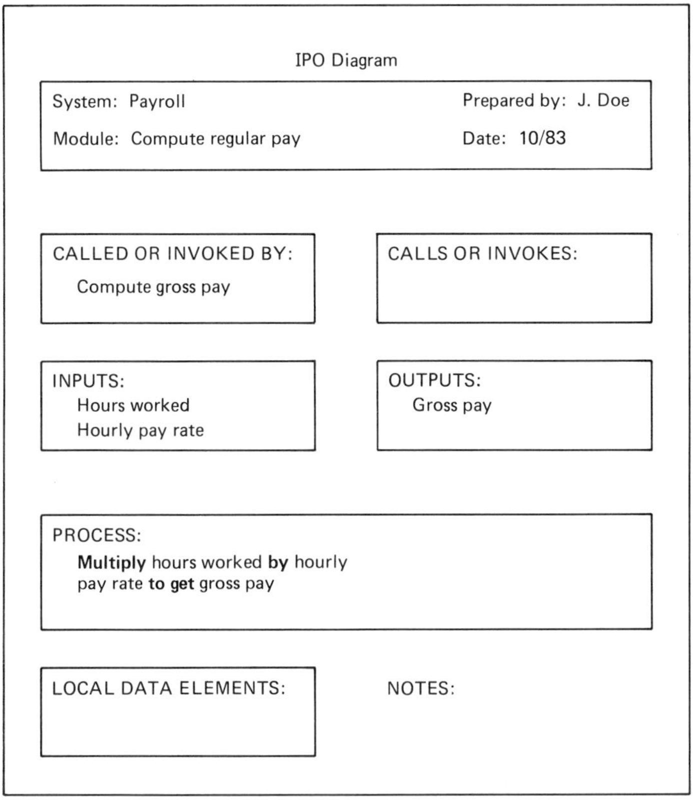

```
┌─────────────────────────────────────────────────────────────────────┐
│                            IPO Diagram                                │
│  ┌───────────────────────────────────────────────────────────────┐  │
│  │  System: Payroll                      Prepared by:  J. Doe     │  │
│  │  Module: Compute regular pay          Date:  10/83            │  │
│  └───────────────────────────────────────────────────────────────┘  │
│                                                                       │
│  ┌───────────────────────────┐    ┌──────────────────────────────┐  │
│  │ CALLED OR INVOKED BY:      │    │ CALLS OR INVOKES:            │  │
│  │   Compute gross pay        │    │                              │  │
│  └───────────────────────────┘    └──────────────────────────────┘  │
│  ┌───────────────────────────┐    ┌──────────────────────────────┐  │
│  │ INPUTS:                    │    │ OUTPUTS:                     │  │
│  │   Hours worked             │    │   Gross pay                  │  │
│  │   Hourly pay rate          │    │                              │  │
│  └───────────────────────────┘    └──────────────────────────────┘  │
│  ┌───────────────────────────────────────────────────────────────┐  │
│  │ PROCESS:                                                       │  │
│  │   Multiply hours worked by hourly                              │  │
│  │   pay rate to get gross pay                                    │  │
│  └───────────────────────────────────────────────────────────────┘  │
│  ┌───────────────────────────────────────────────────────────────┐  │
│  │ LOCAL DATA ELEMENTS:              NOTES:                       │  │
│  │                                                                │  │
│  └───────────────────────────────────────────────────────────────┘  │
└─────────────────────────────────────────────────────────────────────┘
```

Figure 5.10   A different format for HIPO diagrams

SUMMARY

A hierarchy chart, or function chart, shows the hierarchical relations of the modules of a system under consideration.

HIPO is an acronym for Hierarchy plus Input-Process-Output. It is a package that consists of a set of diagrams that graphically describe functions of a system. A tool developed for programming, it is now used for system development and documentation. A typical HIPO package contains three kinds of diagrams: a visual table of contents (or hierarchy diagram), overview diagrams, and detail diagrams. The available aids for preparing HIPO diagrams are the HIPO worksheet and the HIPO template.

Although it is a practical tool for relatively small systems, HIPO becomes impractical for very large systems.

The initial design package, a detail design package, and a maintenance package are three kinds of HIPO packages. The use of HIPO has been reported with some other structured tools such as top-down development, structured programming, chief programmer team, and structured walk-through.

EXERCISES

1. What is the difference between a hierarchy chart and a function chart?
2. For what do the letters ''HIPO'' stand?
3. What are the major objectives of HIPO?
4. What type of diagrams are contained in a HIPO package?
5. What is the rippling effect? How good is HIPO against the rippling effect?
6. Give the names of HIPO documentation packages.
7. What is the relationship of HIPO to some other tools such as top-down development, structured programming, chief programmer team, and structured walk-through?
8. Considering a company that you are familiar with, prepare HIPO packages for
    a. Daily sales
    b. Daily accounts payable
    c. Daily cash receipts procedure
    d. Daily inventory procedure
9. Consider the daily sales of a small trading company. A sales operation for a customer consists of the following major functions:
    a. check customer credit
    b. check goods inventory
    c. process order
    d. print shipping order
    Prepare a HIPO package to represent the daily sales of the company.
10. ABC is a regional airline company. Daily operations of ABC are summarized as ticket sales, plane movements, and personnel operations. Ticket sales consist of local flights and international flights. Plane movements include daily plane landings and plane take-offs. Personnel operations mainly consist of checking the attendance of the personnel. Prepare a HIPO package to represent the daily operations of ABC airlines.

SELECTED REFERENCES

(Au)   Auerbach Publishers Inc. *HIPO*, Portfolio No.: 32-04-06.

(DaW 83)   Davis, W. S. *Systems Analysis and Design*, Addison Wesley, 1983.

(IBM 75)   IBM. ''HIPO—A Design Aid and Documentation Technique,'' Publication No. GC20-1851-1, May 1975.

(Ka 76)   Katzan, H. *Systems Design and Documentation*. Van Nostrand Reinhold, 1976.

(Tu 80)   Turner, J. ''The Structure of Modular Programs,'' *Comm. ACM*, Vol. 23, No. 5 (May 1980), 272–77.

Chapter 6

Data Flow Diagrams

6.1 GENERAL

The idea of representing the flow of data in a system by a chart is not new. As reported by Peterson (Pen 77), in 1962 Petri used circles (called places) and bars (called transitions) to represent the static properties of a system. The nodes, or circles, and bars are connected to each other by directed arcs. In 1967, Martin and Estrin (ME 67) represented computational algorithms by directed graph which had circles and arrows as well as some Boolean operator symbols.

As noted by Gane and Sarson (GS 79, p. 25), it is a relatively new insight to use data flow diagrams at the logical level as a key tool for understanding and working with a system of any complexity, as well as to refine the model for use in analysis. The use of such a tool was first reported by Stevens and his colleagues (SMC 74), although many others later used it under the same name, under the acronym DFD, or as a bubble chart. It has also been used in conjunction with the composite design/structured design methodology of the structured approach that will be discussed in Chapter 15. As we shall see later, the data flow diagram concept has also been used in another methodology, namely in SADT (e.g., RB 76), although the circles are replaced by rectangles.

Data flow diagrams enable us to describe an existing system or a proposed new system at a logical level without considering the physical environment in which data flows (e.g., telephone calls, mail, and so on) or the physical environment in which data is stored (e.g., card file, microfiche, disk, floppy disk, or tape).

6.2 A SIMPLE EXAMPLE

To demonstrate DFD and its various symbols, let us consider the daily operations
of a local flower shop, Angora Florists, Inc. It is known that almost three quarters
of Angora's business is transacted over the phone, mainly for home and/or work
delivery. Customer accounts are maintained in the shop.

Since the customer both delivers the order and receives the flowers plus in-
voices, the customer is both the source in DFD terminology and the destination or
sink. As shown in Figure 6.1, a source or sink is represented by a square in a
DFD. The circle at the middle of the figure is the process bubble, the hub of the
system, so to say. The open-ended rectangles of Figure 6.1 indicate the storage of
goods or data. Thus during the processing of an order, some of the flowers are
retrieved from the store. Customer information is also stored in the shop, and it
passes back and forth between the file and the processing station in order to access
customer address and credit status.

As can be seen in Figure 6.1, there are circle(s), square(s), open-ended rect-
angle(s) and directed arrows in a DFD. In the following section these symbols are
formally defined.

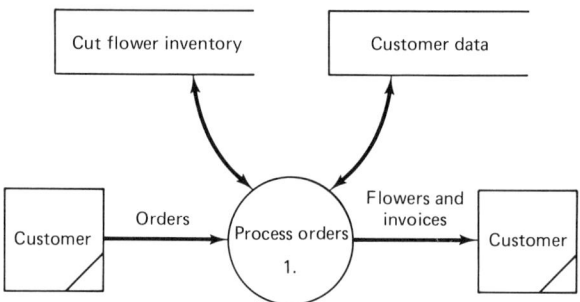

Figure 6.1   Operations of Angora Florists, Inc.

6.3 DFD SYMBOLS

One may define a DFD as a logical model that describes a system as a network of
processes (subsystems) connected to each other and/or to data stores and also to
source(s) and sink(s) (or destinations). It is now proper to define the basic symbols
that are used in a DFD to signify various entities and functions.

1. Data Flow
    An arrow is used to represent a flow of data (information
    or objects). The name of the data flow is written through
    or next to the line.

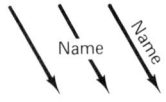

2. Process

A circle (or bubble) represents an automated or manual task or process. It represents not only the input data that flows into the bubble, but also the transformation of that input data into output, which then flows out of the bubble. A brief descriptive statement and a reference number for the process is written inside the bubble. The descriptive statement should be an imperative sentence that consists of an active verb (e.g., verify, compute) followed by an object clause to describe the process explicitly.

3. Edge of the Model or Source/Sink

A square is used as an external entity symbol to represent an area in which data originates (i.e., source) or terminates (i.e., sink or destination). In modelling a system by DFD, the data flows that enter and leave the system define the boundary of the system. The name of the originator/terminator entity is written inside the square box, in singular. A more general definition of source and sink may be as follows: A source is an independent system that produces data flows which our system then processes. Similarly, then, sink is another independent system which receives the data flows that our system produces. Thus, a system can be both a source and a sink, and we are allowed to duplicate the symbol if the same outside system appears as both source and sink on a data flow diagram.

4. Storage

An open-ended rectangle represents a store of information or objects, irrespective of the physical storage medium. The name of the store is written inside the symbol and it should be chosen to be most descriptive to the user. The store symbol identifies a time delay for its content. Quite often data elements do not flow from one process to the next directly but are delayed—that is, stored—while other operations or reordering of data elements occurs. The open-ended rectangle enables us to show such delays in DFD.

If a data store is only updated during or after a certain process, then an arrow from that process leading to that store is used to indicate the data flow into the store. If, however, the data from the store is used in the process, then one should use a bidirectional arrow. If the name of data store defines clearly the incoming and/or outgoing data, it is not necessary to name the data flows. If, however, the incoming and/or outgoing data are only a portion of the data elements of the stored data, one then should name the data flow(s) properly.

5. Naming

The names that are used for data flows, description of processes, source/sink, and storage must be defined in the Data Dictionary (DD) of the system development work. DD will be discussed in Chapter 11.

6. Additional Conventions

The crossing of data flow lines should be minimized. In doing this, the same external entity and/or data store can be drawn more than once on the same DFD. The duplicate external entities (source/sink) are identified by an inclined line (/) or by an asterisk (*). The duplicate data stores are identified by a vertical line (|) or by an astrisk (*). If we have more than one type of duplicated entities, one then uses two or more duplication symbols to indicate that. Figures 6.2 and 6.3 show the symbol duplications.

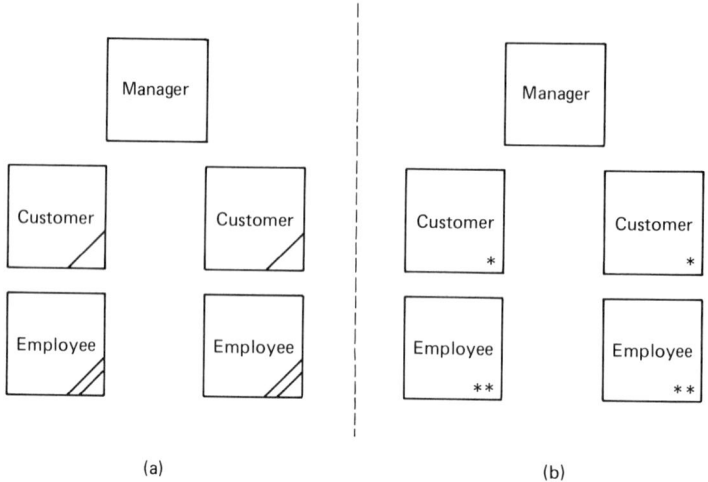

(a)   (b)

Figure 6.2   Example of external entity symbol duplications

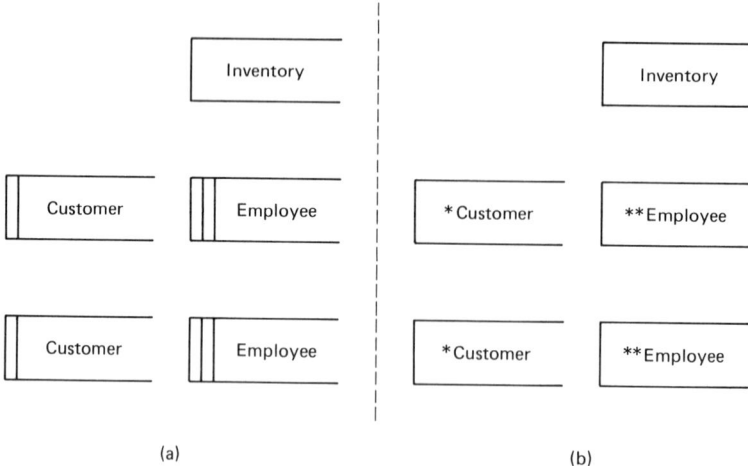

(a)   (b)

Figure 6.3   Example of data store symbol duplications

Customer and employee are the duplicated external entities in Fig. 6.2. Considering customer and employee as data stores, they are shown as duplicated data stores in Figure 6.3.

As shown in Figure 6.4, a little loop is used when it is inevitable that one data flow has to cross another.

Figure 6.4   Data flow line crossings

7. Summary of DFD Symbols
The basic DFD symbols that we have discussed may be summarized for a hypothetical system and presented as Figure 6.5.

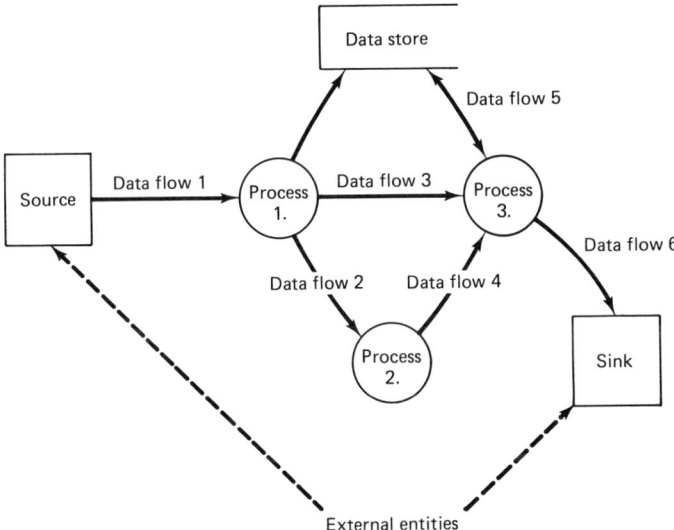

Figure 6.5   Summary of DFD symbols

6.4 OTHER SYMBOLS

Although the symbols defined earlier in this chapter are the most commonly used ones, there are some other symbols in the literature. Several of these are summarized in Figure 6.6. These symbols are given here only for general reference and will not be used in the text.

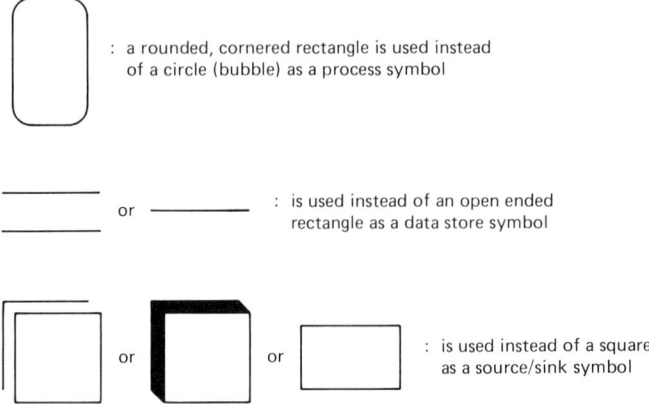

Figure 6.6   Additional DFD symbols

6.5 LEVELS OF DATA FLOW DIAGRAMS

In preparing a DFD the systems analyst tries to identify the typical flows of data and the proper processes that transform them. They are then summarized as a pack of Data Flow Diagrams. In preparing a DFD, the systems analyst uses a top-down approach so that the top page or the first page of the diagram pack shows the boundaries of the system under consideration. This top page is called a context diagram. The major processes of the system and the relevant data flows are described on the next lower level diagram, which is known as an overview diagram or zero-level diagram. On any one page, the 7 ± 2 rule is applied to the number of processes described. Any process on a higher level DFD is further decomposed into an entire new page of DFD on lower level pages, such that incoming and outgoing data flows to the process are also shown on the diagram. Just as processes are decomposed, data flows on DFDs may also be decomposed into their elements on lower level diagrams. Going upward in DFD levels, one then ensures that detailed processes and/or data flows are summed up in the higher level diagrams. The decomposition or explosion process of data flow diagrams is described in Figure 6.7. For example, an explosion of bubble 3 is shown in the level-1 diagram, and an explosion of bubble 3.1 is shown in the level-2 diagram.

6.6 GUIDELINES FOR DRAWING DATA FLOW DIAGRAMS

The basic steps of preparing data flow diagrams are given by various authors (e.g., DeM 78, GS 79, Jo 80, Wei 80, and YC 79). A guideline for drawing data flow diagrams is summarized below:

Step 1:   Identify all the sources and destinations of the system.

Step 2:   Identify all outputs and inputs of the system and draw the context diagram.

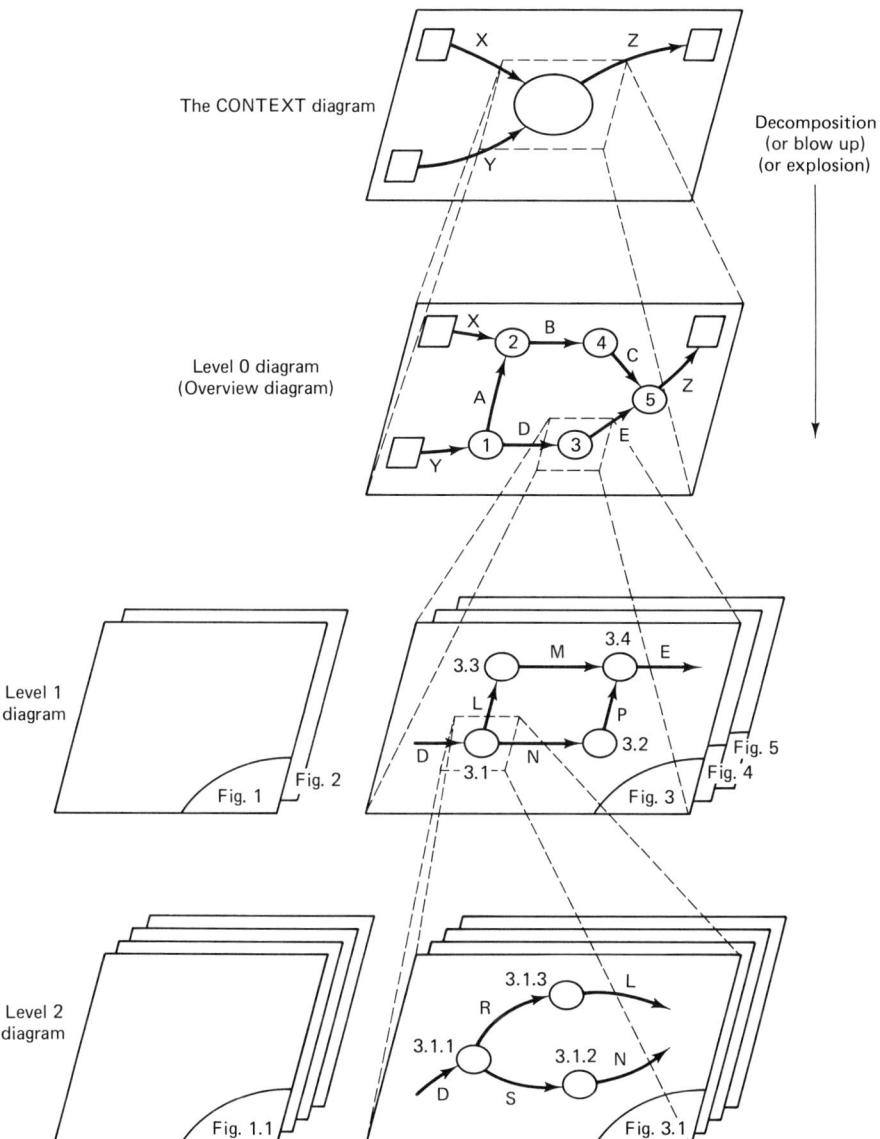

Figure 6.7   Levels of data flow diagrams (DFD)

Step 3:   Taking each source on the left-hand side of the paper, draw the data flows, processes, and data stores that will be needed to arrive at the proper destination(s) on the right-hand side of the paper. When you reach a standstill, you may then take the destinations first and come to the middle of the diagram; in some cases, you may even work your way from the middle—that is, the processes of the diagram—towards the sources and/or destinations.

Step 4:   Label all data flows and data stores to indicate their composition and include them in the Data Dictionary (DD).

Step 5:   Label all processes or transformations (i.e., bubbles), if possible, by means of a transitive verb and an object which is also included in the DD. In labelling processes, consider their inputs and outputs.

Step 6:   Draw the first draft of the diagram free hand, including everything needed except error checking, exception analysis, decision making, initialization (e.g., OPEN FILE), and termination (e.g., CLOSE FILE).

Step 7:   When you are finished with the first draft, check over your list of inputs and outputs that were developed in Step 2. If it is necessary, go over the diagram again. Remember, one may need several drafts.

Step 8:   Next prepare a neater draft using a template and/or other tools. You may use duplicates of data stores and external entities if it helps you to minimize crossing of the data flows on the diagram.

Step 9:   If possible, conduct a review on the revised draft with a user and/or colleague to determine whether the diagram truly represents the system under consideration. If needed, make the necessary modifications.

Step 10:   For each process defined in the final revision of the DFD, work out a lower level decomposition or explosion. Use the proper numbering for parent and children diagrams. If necessary, go back and make additional changes and corrections in the parent DFD.

Step 11:   Repeat the previous step for each process in the DFD until each process has been defined in sufficient detail or in terms of its most elementary inputs and outputs.

6.7 LIMITATIONS AND ADDITIONAL CAPABILITIES OF DFD

As stated earlier, a DFD shows

- partitioning of the systems into subsystems
- data flows in the system
- data stores and in-flowing and out-flowing data
- external entities, that is, sources and sinks of the system

On the other hand, a DFD normally does not show

- composition of data flowing in the system
- data access requirements of data stores
- decisions in the system
- loops in the system
- calculations
- quantities for data and/or processes

It is, however, possible that by defining some relational operators, the capabilities of a DFD can be improved (e.g., Wei 80, YC 79). These operators are:

* denotes a logical AND connection (both a and b);

⊕ denotes an exclusive OR (or XOR) connection (only a or b)

○ denotes an inclusive OR connection (a or b or both).

If both AND and OR operators appear together, the AND operator (i.e.,*) will be performed first and then the OR operator (i.e., ⊕ or ○). An example of how these symbols operate is shown in Figure 6.8. The inclusive OR connection is used for transactions 1 through 4. After editing, a transaction is either valid or rejected. Hence an exclusive OR is used. After an update operation, both a new master file and a list of updates are obtained; hence the AND operator is used.

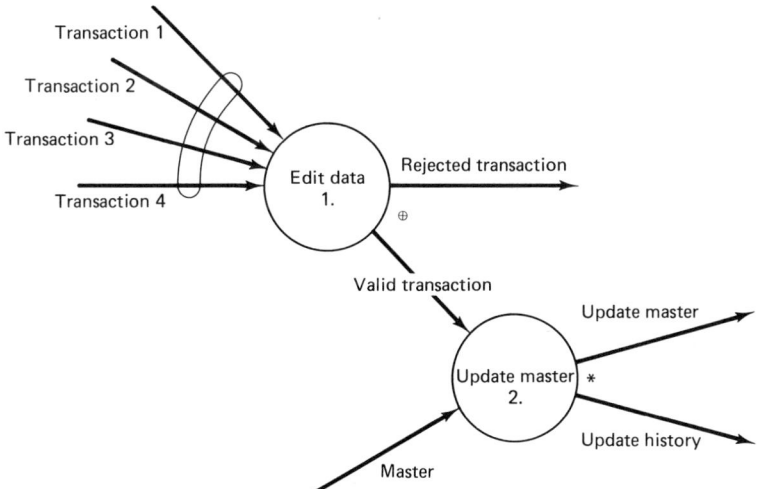

Figure 6.8   Example of logical operator symbols on a DFD

6.8 AN EXAMPLE OF DFD

As an example of drawing a DFD, consider the activities of Hoosier Feed, Inc.
Hoosier Feed receives orders from local farmers for cattle and hog feed. It orders
bulk feed from Happy Mill and breaks the bulk shipment down for individual
farmers. A context diagram and an overview diagram of the activities of this company are given in Figures 6.9 and 6.10, respectively.

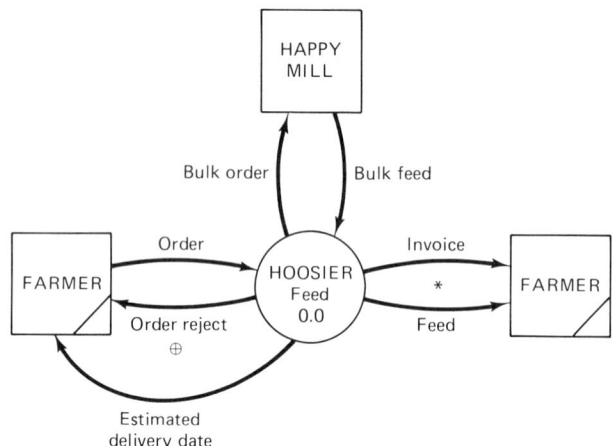

Figure 6.9   Context diagram for the activities of Hoosier Feed, Inc.

6.9 COMMON ERRORS IN DATA FLOW DIAGRAMS

The most common errors in using DFDs may arise from the misuse of several
symbols that have been defined earlier for external entities, stores, and so on, or
they may result from missing and/or extra data flowing into or out of a process
bubble. Obviously, data needed to produce a certain data outflowing from a bubble
must be inflowing to the bubble for processing.

6.10 MAJOR REASONS FOR USING DATA FLOW
DIAGRAMS

The major reasons for using DFDs may be summarized as follows (StW 82):

1. They help systems analysts
   - to summarize information about the system
   - to understand the key components of the system and to define reusable
     functions

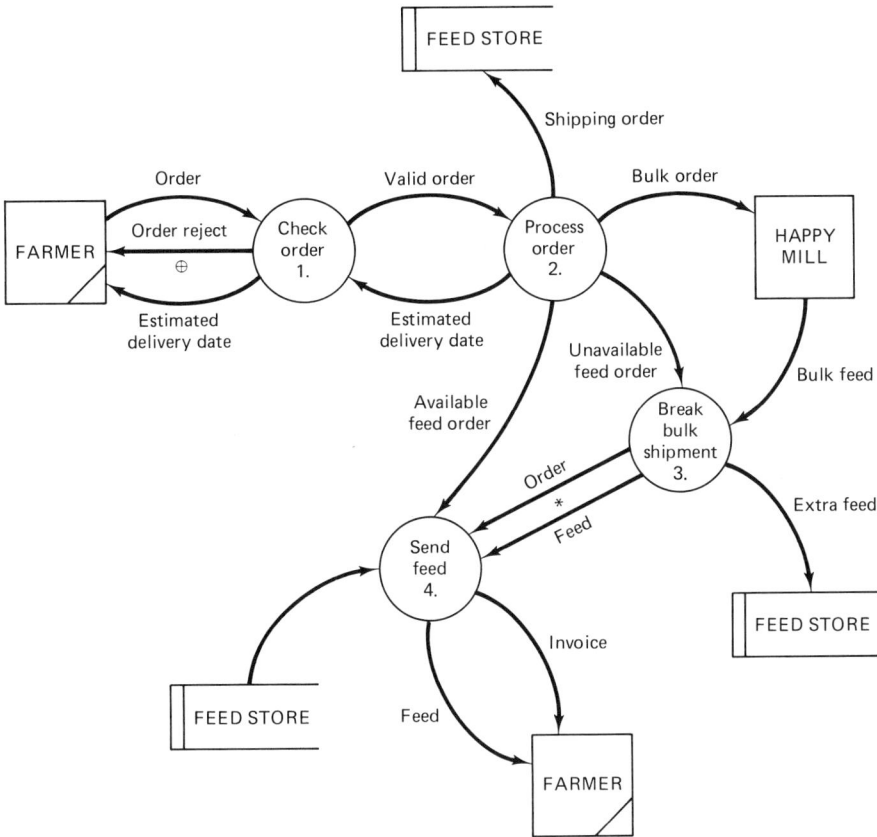

Figure 6.10 Overview diagram for Hoosier Feed, Inc.

- to understand the relationship between subsystems and subassemblies
- to effectively carry through development of the application

Each data flows and the contents of every store in the DFD should be defined in a Data Dictionary (DD), which serves as a first step in developing a database of the system.

2. Recall the fact that communication between user(s) and system analyst(s) is vital in the development of an information system. A DFD serves as an excellent communication tool between these two groups by reducing complexity and by providing a kind of ongoing review of the application.

3. Examining the timing requirements of the various processes in the DFD as a guide, the system analyst is able to draw a number of automation boundaries to develop physical system alternatives. Opposite to decomposition, an automation boundary shows the integration of a group of DFD bubbles into a single process.

SUMMARY

A data flow diagram, DFD, or bubble chart is a logical model to describe an existing system or a proposed new system. A circle is used for process, an arrow for data flow, a square for source/sink, and an open-ended rectangle for data storage. The data names used in a DFD should be defined and their compositions should be described in an accompanying Data Dictionary (DD).

To achieve greater clarity, DFDs are prepared at several levels. The highest level is the ''context diagram,'' which merely shows the boundaries of the system under study. The processes are further decomposed in the lower level diagrams.

A DFD may show the subsystems, data flows, data stores, and external entities of a system. However, it does not show data composition, data access requirements of data stores, decisions and loops of the system processes, calculations, or quantities. Sometimes logical operators such as AND and OR are also used to increase the descriptive capabilities of DFDs.

Common errors in DFDs include misuse of symbols and missing inflowing and/or outflowing data.

Basically DFDs are practical tools to help systems analysts during the analysis and design phase of systems development. One of its major benefits is, however, the fact that it facilitates communication between users and systems analysts.

EXERCISES

1. What is wrong with the following DFDs? Correct them where possible.

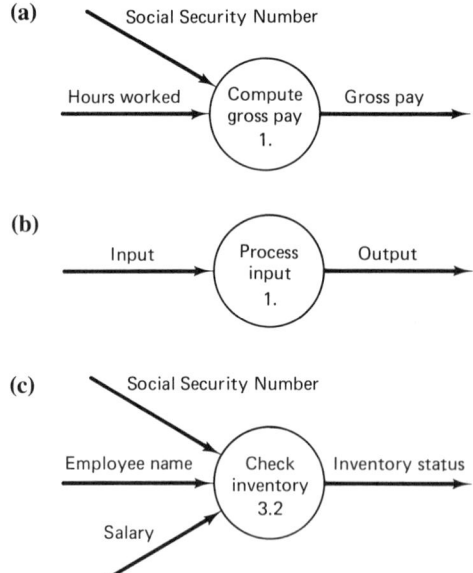

(d)

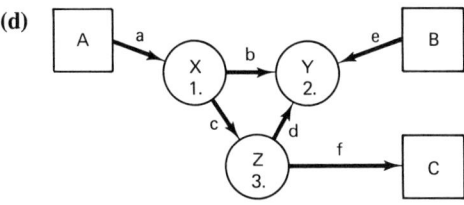

(e)

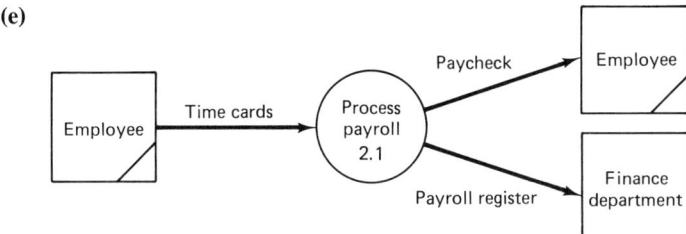

(f)

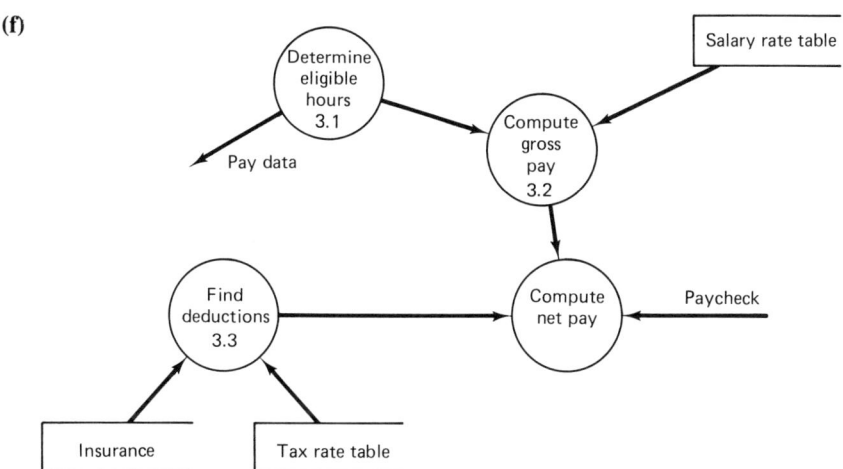

2. Prepare a DFD to describe the activities for
   (a) getting a plane ticket through a travel agent
   (b) eating a doughnut
   (c) pumping fuel into your car
   (d) APM (Auto Parts by Mail) Corporation that stocks auto parts and supplies them by phone and/or mail order
   (e) any other activity of your choice

3. BMB is an international trade corporation. Customer orders are received via phone and telex. Customer order processing, correspondence, and accounting are some of the major processes. Customer file, vendor file, and materials and equipment specifications are some of the data stores in the company. Using DFD, represent the activities of this company.

4. Represent activities of the reserved books section of a university library using DFD.

5. Using DFD, represent the activities for getting a student transcript in a university registration office.

6. Using DFD, represent the various stages of a linear programming (LP) application for a business problem.

SELECTED REFERENCES

(DeM 78)   De Marco, T. *Structured Analysis and System Specification*. Yourdon Press, 1978.

(GS 79)   Gane, C., and T. Sarson. *Structure Systems Analysis: Tools and Techniques*. Prentice-Hall, 1979.

(Jo 80)   Jones, M. P. *The Practical Guide to Structured Systems Design*. Yourdon Press, 1980.

(ME 67)   Martin, D., and G. Estrin. "Models of Computations and Systems," *Journal of the ACM*, Vol. 14, No. 2 (April 1967).

(Pen 77)   Peterson, J. L. "Petri Nets," *ACM Computing Surveys*, Vol. 9, No. 3 (September 1977) 223–52.

(RB 76)   Ross, D. T., and J. W. Brackett. "An Approach to Structured Analysis," *Computer Decisions*, Vol. 8, No. 9 (September 1976), 40–44.

(SMC 74)   Stevens, W. P., G. J. Myers, and L. L. Constantine, "Structured Design," *IBM Systems Journal*, Vol. 13, No. 2 (1974), 115–39.

(StW 82)   Stevens, W. P. "How Data Flow Can Improve Application Development Productivity," *IBM Systems Journal*, Vol. 21, No. 2 (1982), 162–78.

(Wei 80)   Weinberg, V. *Structured Analysis*. Prentice-Hall, 1980.

(YC 79)   Yourdon, E., and L. L. Constantine. *Structured Design*. Prentice-Hall, 1979.

Chapter 7

SADT

7.1 GENERAL

SADT is an acronym for *S*tructured *A*nalysis and *D*esign *T*echnique, a tool that was developed by D. T. Ross during the period from 1969 to 1973. It is now supported and marketed by SofTech, Inc. as a tool that can be used in all phases of systems development. Ross and his colleagues described SADT as both a graphic language for describing systems and a methodology for producing such descriptions (RDMcG). A system is viewed as consisting of things (objects, documents, or data), happenings (activities performed by people, machines, or software), and their interrelationships in SADT applications.

There are two types of diagrams that are used in a SADT package:

1. Activity diagrams (called actigrams)
2. Data diagrams (called datagrams)

Actigrams and datagrams are organized separately in a top-down manner; each diagram is either a summary (i.e., parent) diagram or a detail (i.e., child) diagram of the parent.

The activity kit of SADT consists of activity diagrams that are oriented toward exploding the activities of the system. The data kit, on the other hand, consists of data diagrams oriented toward the depiction of data decomposition in the system. The activity and data kits of SADT are also termed models in the literature. Each type of model includes both data and activities; the main difference is on the subject of decomposition. Any SADT diagram is made up of 7 ± 2 boxes and the arrows connecting these boxes.

A special case of SADT is reported by Peters (Pes 81) as SAMM (Systematic Activity Modelling Method) in which only the activity diagrams are utilized.

7.2 SADT ACTIVITY BOX

On an activity diagram of an activity kit (or activity model) of SADT, the boxes correspond to activities and the arrows correspond to data as can be seen in Figure 7.1.

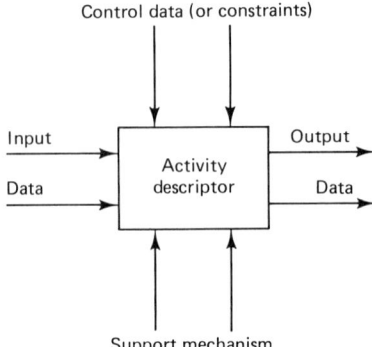

Figure 7.1   SADT activity box

Each activity represented by a box is named with a verb or verb phrase and an object; an example would be ''sort employee data.'' The phrase ''control data'' as applied to such a box expresses the constraints on that activity. The supporting mechanism of the activity by contrast identifies the department or individual related to and/or responsible for the activity. The support mechanism is also used for cross-referencing models. An example of an SADT activity diagram is given in Figure 7.2. In that figure there are three activities: compute gross pay, compute net pay, and print paycheck. On the figure the data and constraints of the activity boxes are also given.

7.3 SADT DATA BOX

The rectangular boxes in data diagrams of a data kit (or data model) of SADT correspond to data, and the arrows on the diagram indicate the activities related to

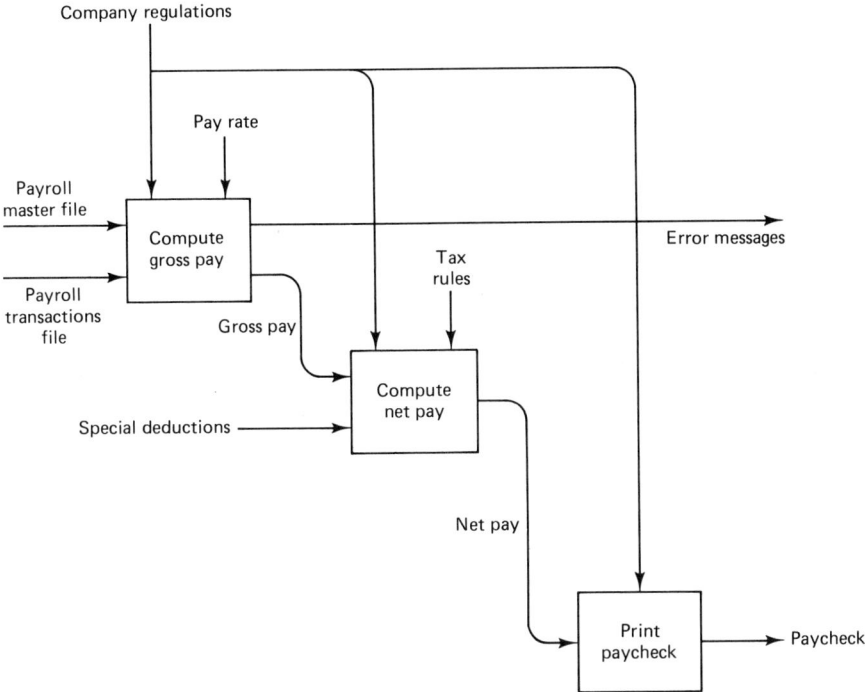

Figure 7.2   Example of an activity diagram

that data, as shown in Figure 7.3. A box on a data diagram is named with a noun or a noun phrase. The control activity at the top of the box is the activity that limits the generation and use of the data. In this case, the box is supported by the storage mechanism (e.g., file), which also serves as a way to cross-reference models, just

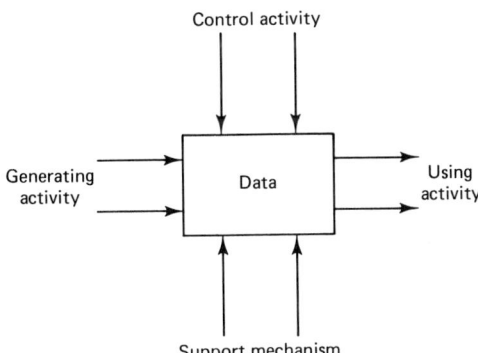

Figure 7.3   SADT data box

as in an activity model. The "generating activity" refers to the activity that generates data, and the "using activity" to the activity that uses data.

7.4 AN EXAMPLE

The sales commission of a salesperson in ABC Co. is computed using the person's rank and monthly total sales. In computing the commission, company policy and company profit for the last six months should be used as control data. The accounting department is responsible for commission computation. Using this information, one can prepare an activity box and a data box of SADT as shown in Figure 7.4a and b, respectively.

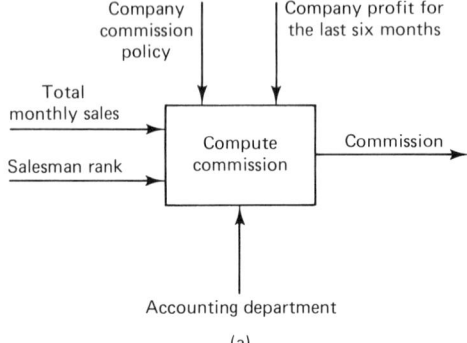

(a)

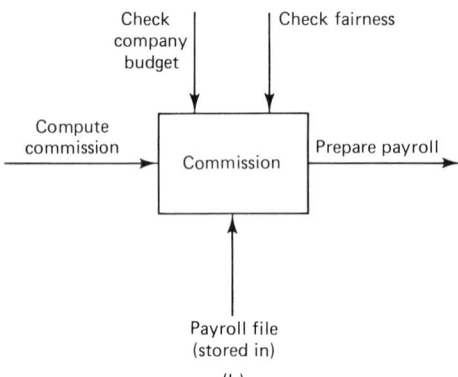

(b)

Figure 7.4 (a) Example activity box, (b) Example data box

7.5 DECOMPOSITION IN SADT

Activity diagrams and data diagrams of SADT are decomposed in the same way that data flow diagrams were exploded in Chapter 6. Figure 7.5 reflects the de-

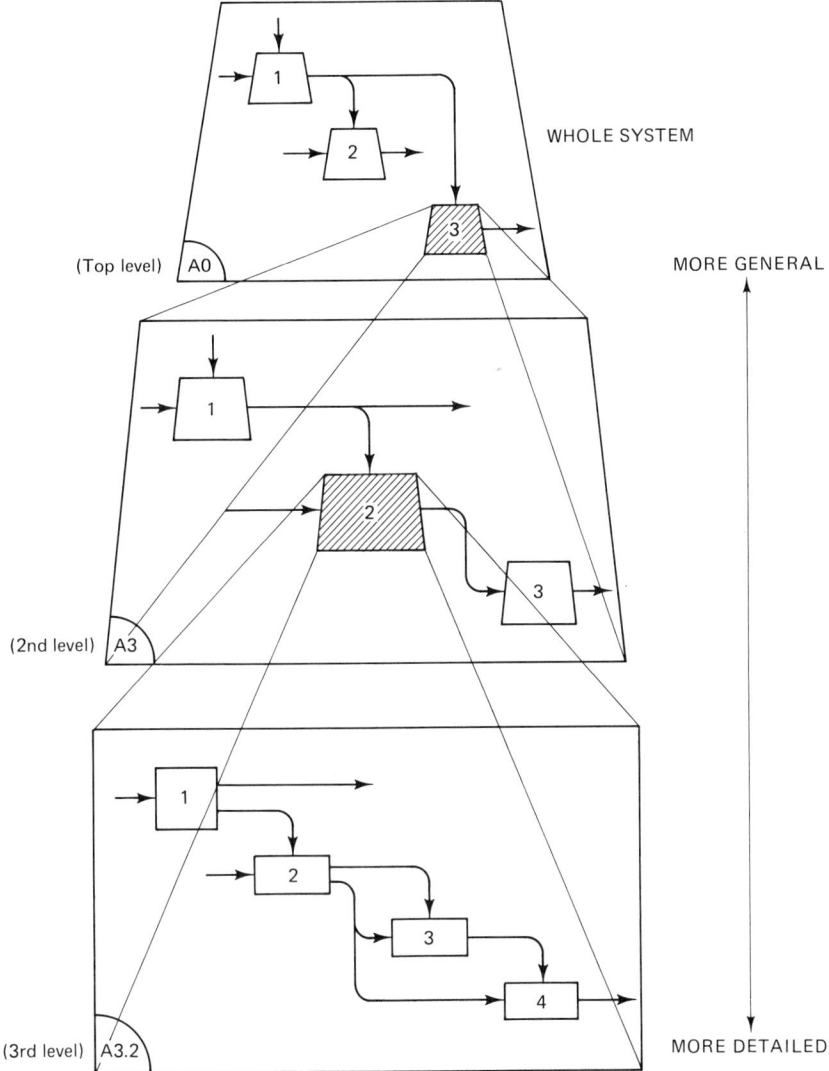

WHOLE SYSTEM

MORE GENERAL

(Top level) A0

(2nd level) A3

(3rd level) A3.2

MORE DETAILED

Figure 7.5   Decomposition in SADT

composition of an activity diagram of SADT. An example of SADT in action will be given in Chapter 15.

7.6 GENERAL REMARKS ABOUT SADT

The separate representation of data and activities within the information system, indicating at the same time, however, their links, is the strength of SADT. The major disadvantage of SADT is stated by Peters (Pes 81) as its richness: An SADT diagram contains so much information, it may be hard for users to appreciate it as a proper and correct model for their systems. Further material about SADT may be found in Connor (Co 80).

SUMMARY

As its name implies, SADT (Structured Analysis and Design Technique) is a tool that can be used for both the analysis and the design phases of information systems development. An SADT package consists of two types of diagrams—activity diagrams and data diagrams. Both types of diagrams are organized separately in a top-down manner that recalls the decomposition of data flow diagrams.

An activity diagram of SADT shows the activities, their relations, and their decompositions. A data diagram, on the other hand, is oriented toward the depiction of the decomposition of the data in the system.

EXERCISES

1. What does SADT stand for?
2. What types of diagrams are there in an SADT package?
3. What is the difference between an actigram and a datagram?
4. What is the difference between SAMM and SADT?
5. Prepare an activity diagram and a data diagram of a small information system that you are familiar with.
6. The following DFD is given:

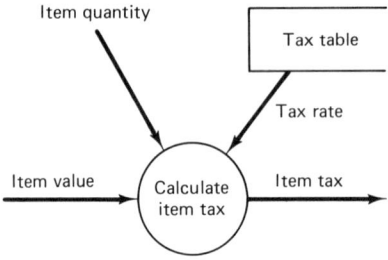

Convert it into a proper SADT diagram.
7. Prepare an activity diagram for a telephone billing project.
8. Prepare the activity diagram and data diagram of the customer billing system of a company.
9. Prepare the activity diagram and data diagram of a student registration system.

10. Prepare the activity diagram and data diagram for the forecasting activities in a production system.

11. Represent an information retrieval system using SADT diagrams.

SELECTED REFERENCES

(Co 80)   Connor, M. F. *Structured Analysis and Design Technique.* SofTech, Inc., May 1980.

(FW 80)   Freeman, P., and A. I. Wasserman, *Tutorial on Software Design Techniques*, 3rd ed. IEEE Computer Society, 1980.

(Pes 81)   Peters, L. J. *Software Design.* Yourdon Press, 1981.

(RDMcG)   Ross, D. T., M. E. Dickover, and C. McGowan. *Software Design Using SADT.* Auerbach, Inc., Portfolio No: 35-05-03.

(RS 80)   Ross, D. T., and K. E. Schoman. "Structured Analysis for Requirements Definition," in *Tutorial on Software Design Techniques*, ed. P. Freeman and A. I. Wasserman. IEEE Computer Society, 1980, 97–106.

(Ro 80)   Ross, D. T. "Structured Analysis (SA): A Language for Communicating Ideas," in *Tutorial on Software Design Techniques*, ed. P. Freeman and A. I. Wasserman. IEEE Computer Society, 1980, 107–125.

Chapter 8

Structure Charts

8.1 GENERAL

Like a hierarchy chart, a structure chart reveals both the modular structure of a system (i.e., it partitions into modules) and the hierarchy into which the modules are arranged. In addition, a structure chart also shows the data and control interfaces among modules. It does not, however, show the decision structure of a system except the major decision(s). Each module in a structure chart is represented by a rectangle identified by a functional module name. The module name consists of a descriptive verb and a single, nonplural object (Figure 8.1a).

Connections and couples are the other two basic elements of structure charts. A connection is represented by a vector joining two modules, and it indicates any reference from one module to something defined in another module. Usually it is used to call other module(s) from a module. A couple is represented by a short arrow with a circular tail. A couple is used to indicate a data item or a control element that moves from one module to another. Data communication is shown by an arrow with an open circle, and control communication by an arrow with a solid circle (Figure 8.1b and c).

In addition to the above elements, a semicircle and a diamond shape are used to represent a major loop and a major decision in a module, respectively. These symbols should be used to represent only the most critical loop(s) and decision(s) in the system to minimize the crowding of the structure chart (Figure 8.1d and e).

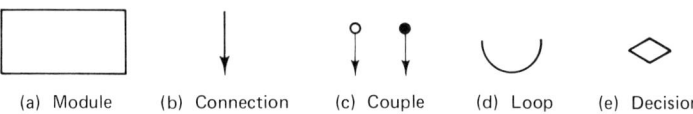

Figure 8.1   Structure chart symbols

8.2 STRUCTURE CHART SYMBOLS AND AN EXAMPLE

The elements of a structure chart—namely, modules, connections, and couples—and their respective symbols are summarized as Figure 8.2.

Figure 8.3 shows an example of a payroll preparation function where em-

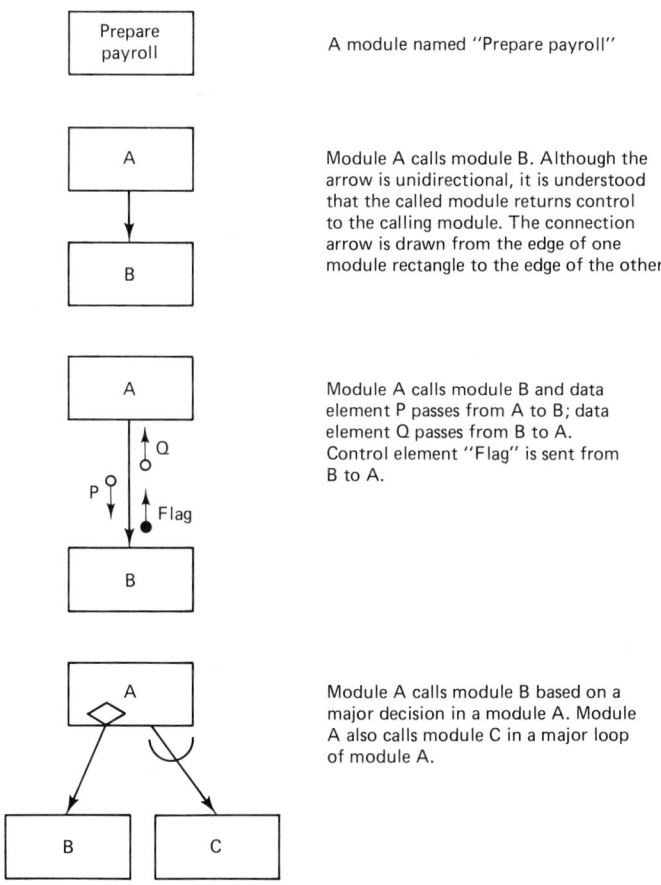

Figure 8.2   Summary of structure chart symbols

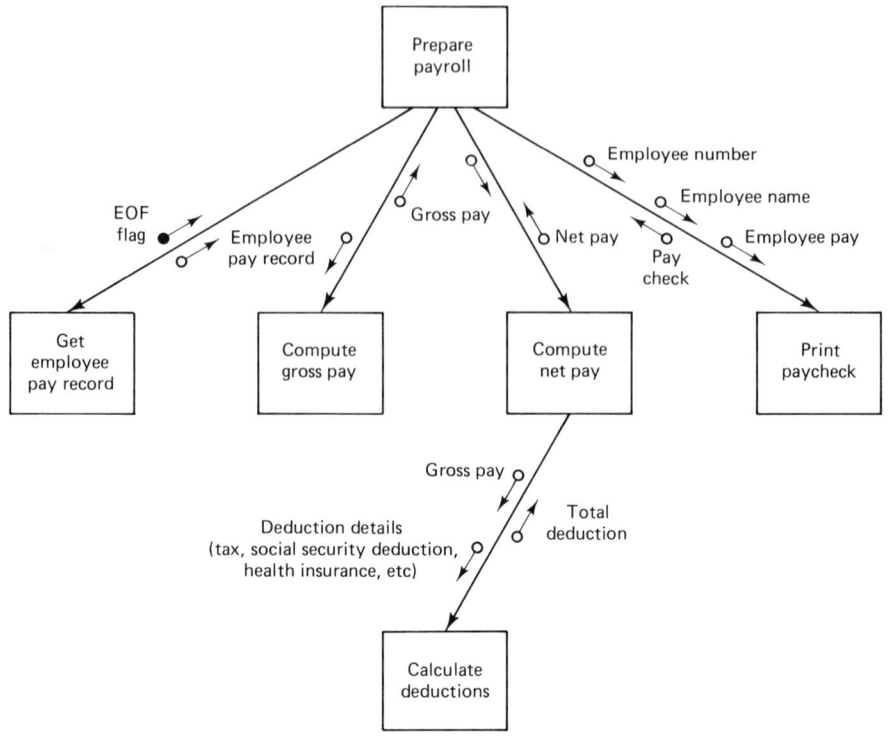

Figure 8.3   Structure chart example

ployee pay record retrieval, gross pay computation, net pay computation, and pay-check printing are considered to be the main modules.

8.3 FINAL REMARKS ABOUT STRUCTURE CHARTS

Structure charts not only illustrate hierarchical modular structures but also indicate the module functions and the data and control communications between modules. They also indicate the fact that higher level modules will be called before lower level ones. As noted by Peters (Pes 80), although structure charts do not permit the documentation of detailed decision information, the sequence of execution is implied in reading from left to right. De Marco (DeM 78), Hassel and Law (HL 82), M. P. Jones (Jo 80), Weinberg (Wei 80) and Yourdon and Constantine (YC 79) are additional references.

SUMMARY

A structure chart shows the modular structure of a system. Inherent in the concept of modular structure is the partitioning of a system into modules or components, the hierarchy into which the modules are arranged, and the interfaces among them. The basic symbols of a structure chart are rectangles to name modules; arrows to represent connections between modules, and couples as a short arrow with a circular tail to show data and control elements communicated between modules. Rarely some additional symbols, such as a semicircle or diamond shape, are also used to show major loop(s) and major decision(s), respectively.

EXERCISES

1. What is the major reason for using structure charts?
2. What is the difference, if any, between the following modular structures?

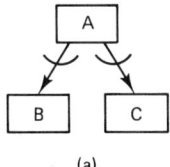 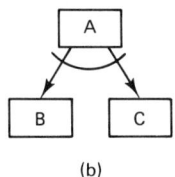

  (a)   (b)

3. You are given the following figure:

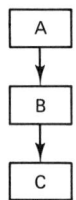

   Is there any difference if you assume this figure is a flowchart over against structure chart?
4. How is a diamond shape used in a structure chart?
5. What is the basic difference between a structure chart and a hierarchy chart?
6. Consider a small farm that produces barley and corn as two major crops. Each year after the harvest, the barley area and corn areas are determined for the following year to optimize the profit of the farm in terms of bushels of yield and selling price of each of the crops.
   Prepare a structure chart for crop planning for this farm by getting the crop record, computing the barley and corn areas, and printing the result.
7. Consider the credit purchase system of a company, and on a structure chart show how to compute and print the monthly schedule of payments resulting from a credit purchase, given the amount of credit purchase and the number of monthly payments planned.
8. Prepare a structure chart for an inventory update operation of a company in terms of the following processes: get transaction, get inventory, process transaction, rewrite inventory, and write reorder.

9. In a government office, birth data is processed to update the birth statistics file. The three major operations during the process are getting the birth record, checking the birth type, and updating a statistics file. The birth type can be normal, or the child is registered to mother only or to father only, or as an orphan child.

Prepare a structure chart to describe the major operations of such an office.

10. One of the current projects of a common carrier company is the editing of the tickets of a telex communication system. The major processes of the tickets editing operation are printing page headings; getting the ticket record; checking the required fields such as call number, country number, sequence number, date and time; storing verified ticket records; and printing an error report.

Prepare a structure chart for such an editing system.

SELECTED REFERENCES

(DeM 78)   DeMarco, T. *Structured Analysis and System Specification.* Yourdon Press, 1978.

(HL 82)   Hassell, J., and V. J. Law. "Tutorial on Structure Charts as an Algorithm Design Tool," *ACM/SIGCSE Bulletin*, Vol. 14, No. 1 (February 1982), 211–23.

(Jo 80)   Jones, M. P. *The Practical Guide to Structured Systems Design.* Yourdon Press, 1980.

(Pes 81)   Peters, L. J. *Software Design.* Yourdon Press, 1981.

(Wei 80)   Weinberg, V. *Structured Analysis.* Prentice-Hall, 1980.

(YC 79)   Yourdon, E., and L. L. Constantine. *Structured Design.* Prentice-Hall, 1979.

Chapter 9

Warnier/Orr Diagrams

9.1 INTRODUCTION

The basic format of the Warnier/Orr Diagrams was developed in the late sixties and early seventies by J. D. Warnier in Paris. It was proposed as a diagram to represent the hierarchical structure of output and input data sets of programs. K. Orr of Topeka, Kansas, later extended some of Warnier's original concepts to information systems analysis and design as well as to database design. The diagrams are, therefore, referred to as Warnier/Orr Diagrams (W/O Diagrams) in the text. They are used to represent data structures as well as processes.

The main tool in a W/O Diagram is the brace "{," also called a universal, which shows decomposition of the system that it depicts. Items that do not decompose further are called elements. If a data structure is represented by a W/O Diagram, the elements are data elements; and if a process of a system is expressed, then its elements are elementary operations.

Other than hierarchy, which is shown in decomposition, the following three simple constructs, representing various data and process structures, can be expressed by W/O Diagrams:

1. sequence
2. repetition (or iteration)
3. selection (or alternation)

In addition, two additional constructs are used as complex constructs:

1. concurrency
2. recursion

The following relational operators are also used in W/O Diagrams:

Symbol	Meaning
$\oplus$	exclusive OR (*a* or *b* but not both)
+	inclusive OR (*a* or *b* or both)
⊘ ⊛ ⊟ ⊞	arithmetic operator
$\overline{\text{process}}$	negation

Concatenation in data structures and in processes is expressed using a dot:

set name { . attribute = set name . attribute

For example, *employee* { . *name* implies *employee name*.

9.2 GENERAL FORM OF SIMPLE STRUCTURES USING W/O DIAGRAMS

Hierarchy. The general form of hierarchy in a W/O Diagram is

$$aaa \{ bb \{ c$$

which means *aaa* consists of *bb* and *bb* consists of *c*.

Sequence. The general form of sequence in a W/O Diagram is

$$aaa \left\{ \begin{array}{l} aa \\ bb \\ cc \end{array} \right.$$

which means *aaa* consists of *aa* followed by *bb* and followed by *cc*. Thus, sequence is represented by listing elements serially within a level of hierarchy.

Repetition. The general form of repetition or iteration in W/O Diagrams is

$$aaa \{ \quad \text{or} \quad aaa \{ \quad \text{or} \quad aaa \{$$
$$(1,N) \qquad (N) \qquad (10)$$
$$\text{or}$$
$$(0,N)$$

$$\text{(i)} \qquad \text{(ii)} \qquad \text{(iii)}$$

   i. $(1,N)$ indicates that *aaa* occurs one to N times
       DO UNTIL (at least once)
  ii. (N) or $(0,N)$ indicates that *aaa* occurs zero to N times
       DO WHILE (zero time is possible);
 iii. (10) indicates that *aaa* occurs just ten times

In the above forms of (i) and (ii), the value of N is unknown; one time or (1) is also expressed by writing nothing below *aaa*.

Selection (or alternation).   A selection structure is represented by using $(0,1)$ (read as zero or one time) and exclusive OR symbol $\oplus$. The general form is:

$$
aaa \begin{cases} bb\ \{ \\ (0,1) \\ \oplus \\ cc\ \{ \\ (0,1) \end{cases} \quad \text{or} \quad aaa \begin{cases} bb\ \{ \\ (0,1) \\ \oplus \\ \overline{bb}\ \{ \\ (0,1) \end{cases}
$$

9.3 COMPLEX STRUCTURES WITH W/O DIAGRAMS

As noted earlier, concurrency, that is, simultaneous operation, and recursion are also expressed using W/O Diagrams.

Concurrency.   The general form of concurrency is

$$
aaa \begin{cases} bb \\ + \\ cc \end{cases}
$$

aaa consists of both *bb* and *cc* and their order is not important. Clearly then + serves as a concurrency operator.

Recursion.   A recursive function is a function that calls itself. In systems descriptions sometimes the concept of recursion is used. Recursion on a W/O Diagram is shown by a broken brace:

The general form of recursion in W/O Diagrams is:

$$aaa \begin{cases} bb \\ \\ aaa \begin{cases} \\ \\ \\ \end{cases} \end{cases}$$

and it means that *aaa* consists of *bb* and itself. A simple example is

$$system \begin{cases} system \begin{cases} \\ \\ \\ \end{cases} \end{cases}$$

which means a system has subsystems.

9.4 EXAMPLES OF DATA STRUCTURES USING W/O DIAGRAMS

In the following examples, a hierarchy structure is expressed by using more than one brace to show the decomposition. In addition to hierarchy, other structures are also demonstrated.

Sequence.   The following diagram depicts an employee record:

Number		Name	Date of Birth		
Div.	Ser. No.		Day	Month	Year

Such a record is represented by a W/O Diagram as:

$$\text{Employee Record} \begin{cases} \text{Number} \begin{cases} \text{Div.} \\ \text{Ser. No.} \end{cases} \\ \text{Name} \\ \text{Date of Birth} \begin{cases} \text{Day} \\ \text{Month} \\ \text{Year} \end{cases} \end{cases}$$

Repetition.   The repetition structure may be shown by considering an employee file consisting of employee records:

$$
\text{Employee File} \left\{ \underset{(1,N)}{\text{Employee Record}} \left\{ \begin{array}{l} \text{Number} \left\{ \begin{array}{l} \text{Div.} \\ \text{Ser. No.} \end{array} \right. \\[1em] \text{Name} \\[1em] \text{Date of Birth} \left\{ \begin{array}{l} \text{Day} \\ \text{Month} \\ \text{Year} \end{array} \right. \end{array} \right. \right.
$$

Selection (or alternation).   Consider an account balance check in bank operations to illustrate selection:

$$
\text{Balance-check} \left\{ \begin{array}{l} \text{Balance} > 0 \ \{ \text{ payment} \\ \quad (0,1) \\[1em] \oplus \\[1em] \overline{\text{Balance} > 0} \ \{ \text{ print ``overdrafting'' message} \\ \quad (0,1) \end{array} \right.
$$

Concurrency.   Now consider the daily operations of a computer system as an example of concurrency:

$$
\underset{(D)}{\text{Daily Operation}} \left\{ \begin{array}{l} \text{Batch} \left\{ \begin{array}{l} \text{Editing} \\ \text{Priority} \\ \text{Job Accounting Report} \end{array} \right. \\[1em] + \\[1em] \text{On-line} \end{array} \right.
$$

This means batch and on-line operations may occur concurrently.

Recursion.   An assembly problem may be represented using recursion:

$$
\text{Assembly} \left\{ \begin{array}{l} \underset{(1,P)}{\text{Pieces}} \\[1em] + \\[1em] \underset{(A)}{\text{Assembly}} \left\{ \begin{array}{l} \\ \\ \end{array} \right. \end{array} \right.
$$

It means that an assembly is defined in terms of another assembly or assemblies. Similarly a part may be defined in terms of other part(s).

9.5 PROCESS REPRESENTATION BY W/O DIAGRAMS

As noted earlier, W/O Diagrams are used to represent processes as well as data structures. For a representation of a process the following general form is used:

$$
\text{Process} \begin{cases} \text{Beginning of the Process} & \text{or} \quad \text{. Begin} \\ \text{Middle of the Process} & \\ \text{End of the Process} & \text{or} \quad \text{. End} \end{cases}
$$

As an example of a W/O Diagram to represent a process, consider an operation that updates a customer master file with sales transactions:

$$
\text{Update Master File} \begin{cases} \text{. Begin} \\ \text{Customer} \\ \quad \text{(C)} \\ \text{. End} \end{cases}
\begin{cases} \text{. Begin} \\ \text{Get transaction} \\ \text{Get master record} \\ \text{Process transaction} \\ \text{Write new master} \\ \text{Store new master} \\ \text{. End} \end{cases}
\begin{cases} \text{. Begin} \\ \text{Read transaction} \\ \text{Edit transaction} \\ \text{.End} \end{cases}
$$

This diagram also shows the hierarchy of operations.

9.6 EXAMPLES OF PROCESS REPRESENTATION USING W/O DIAGRAMS

A common carrier company provides telex communication service. One of the current projects of the company is the computerization of the billing operation of telex communications services. The major steps of the billing operation are creating and updating the master file, and preparing reports. Using W/O Diagrams, these steps of the billing project may be summarized as follows:

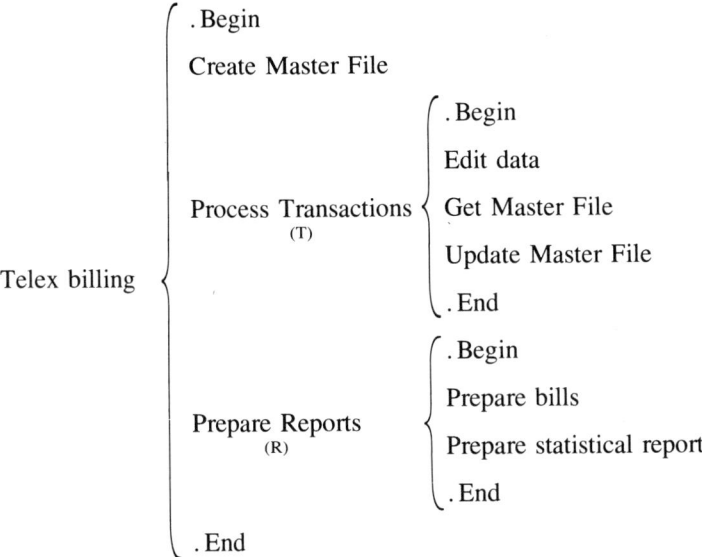

Another example of process representation using W/O Diagrams is the system development life cycle that was discussed in Chapter 2 and presented as Figure 2.1 or as Table 2.2.

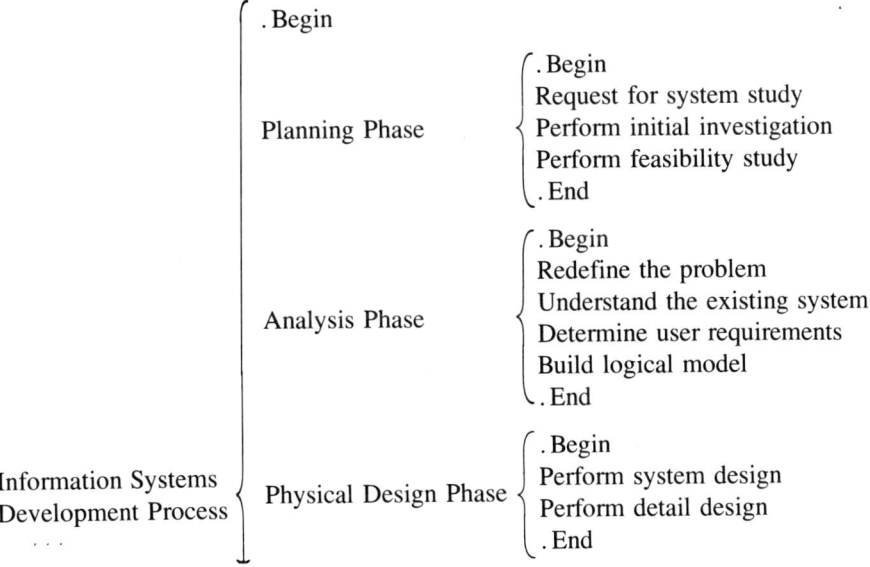

(cont.)

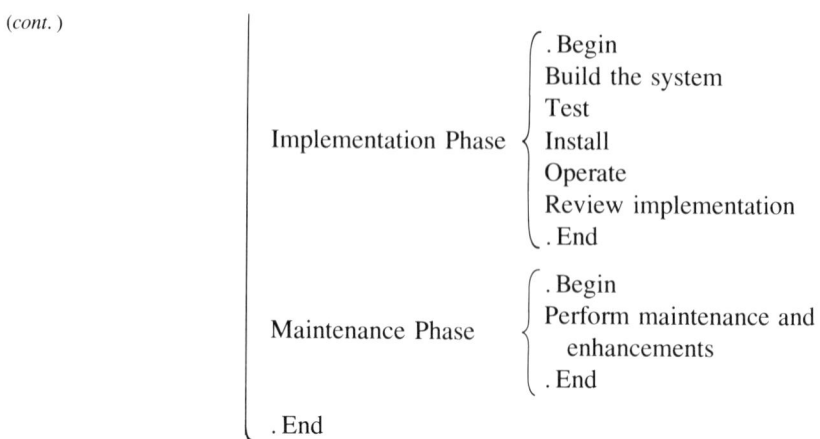

Implementation Phase
- .Begin
- Build the system
- Test
- Install
- Operate
- Review implementation
- .End

Maintenance Phase
- .Begin
- Perform maintenance and enhancements
- .End

.End

9.7 AN EXTENSION OF W/O DIAGRAMS

A combination of systems flow diagram symbols together with W/O Diagram symbols may be used to describe the flow of data in an information system, in particular a database application, where the system consists of three phases: edit, process, and report.

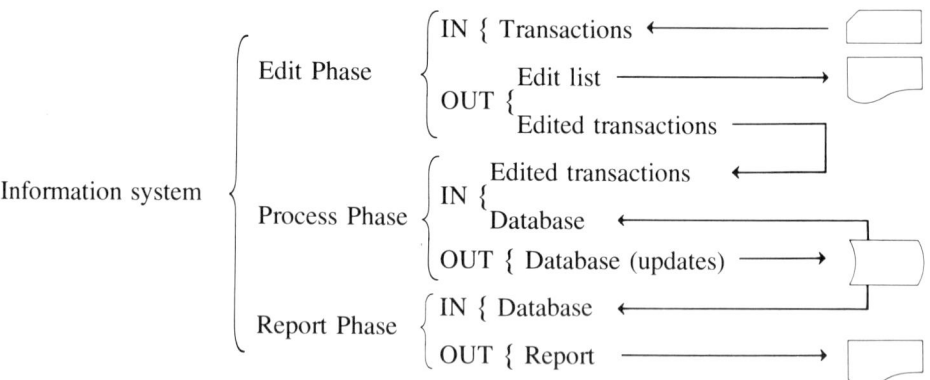

Information system
- Edit Phase
  - IN { Transactions
  - OUT {
    - Edit list
    - Edited transactions
- Process Phase
  - IN {
    - Edited transactions
    - Database
  - OUT { Database (updates)
- Report Phase
  - IN { Database
  - OUT { Report

9.8 FINAL REMARKS ABOUT W/O DIAGRAMS

In addition to the publications of J. D. Warnier (Wa 74 and Wa 81) and K. Orr (Orr 77 and Orr 81b), the major references for Warnier/Orr Diagrams are the publications of Higgins (Hi 79 and Hi 83) and the Auerbach Portfolios, particularly No. 12-03-06. The books published by Higgins emphasize design and construction of programs using W/O Diagrams and W/O methodology that will be included in Chapter 15.

SUMMARY

The main element of a Warnier/Orr (W/O) Diagram is the brace {, which indicates decomposition or hierarchy within the system it describes. In addition to decomposition, W/O Diagrams portray the simple structures of sequence, repetition, and selection (or alternation). Concurrency and recursion are two additional structures that may be represented. Exclusive OR and inclusive OR operators are used as logical operators on a W/O Diagram and an arithmetic operator is set inside a box. Negation is expressed in a W/O Diagram by an overbar.

W/O Diagrams are used to describe data structures, processes, and data flows in an information system. In representing data flows, system flow diagram symbols are also used together with braces.

EXERCISES

1. What is the basic element of a W/O Diagram?
2. Can processes also be represented by means of W/O Diagrams? If so, explain how.
3. What are the basic structures represented by W/O Diagrams?
4. What additional structures are also represented by W/O Diagrams?
5. What type of operations are used in W/O Diagrams?
6. Represent your daily activities by using W/O Diagram(s).
7. Take a hypothetical inventory update process and represent it using W/O Diagrams.
8. Describe the situation using W/O Diagrams if you submit your homework or not.
9. Describe your near future after graduation using W/O Diagrams for the description.
10. Represent a customer record of a local bank using W/O Diagrams.
11. Represent the data structure of the available books in a local public library using W/O diagrams. Using W/O Diagrams, represent a search process on the master file of the library.
12. The personnel record of an organization contains the fields of ID, rank and position, personal data, job history, and salary. The ID field consists of number, name, and address; personal data consists of birth date, marital status, sex, and number of dependents; and job history contains date hired and expected retirement date. Represent the structure of such a record using W/O Diagrams.
13. The vehicle file is one of the available files of a Traffic Information System of a town. Such a file contains technical and registration data of the recorded vechicles. A vehicle record has ID data, technical specifications, registration data, and vehicle history. Represent such a record using W/O Diagrams.
14. Represent a telephone subscription record using W/O Diagrams.
15. Represent the quality control system of an industrial organization using W/O Diagrams.
16. Solve problems 8 through 10 of Chapter 5 using W/O Diagrams.
17. Solve problems 3 through 6 of Chapter 6 using W/O Diagrams.
18. Solve problems 6 through 9 of Chapter 8 using W/O Diagrams.

SELECTED REFERENCES

(Au)   Auerbach, Inc. *Warnier-Orr Diagrams*, Portfolio No: 12-03-06.

(Hi 79)   Higgins, D. *Program Design and Construction.* Prentice-Hall, 1979.

(Hi 83)   Higgins, D. *Designing Structured Programs.* Prentice-Hall, 1983.

(Orr 77)   Orr, K. *Structured Systems Development.* Yourdon Press, 1977.

(Orr 81b)   Orr, K. *Structured Requirements Definition.* K. Orr & Associates, 1981.

(Wa 74)   Warnier, J. D. *Logical Construction of Programs.* Van Nostrand/Reinhold, 1974.

(Wa 81)   Warnier, J. D. *Logical Construction of Systems.* Van Nostrand/Reinhold, 1981.

Chapter 10

Other Diagrams

10.1 INTRODUCTION

In addition to the diagrams discussed in the previous chapters, other diagrams used in structured systems analysis and design include Jackson's Diagrams, Chen's E-R representation, and Leighton diagrams. These tools are defined and discussed in the following sections.

10.2 JACKSON'S NOTATION

Using Jackson's notation, we can represent any program, data structure, or information system in terms of system hierarchy and its elementary and composite components. Elementary components are those that are not further decomposed and have no parts. Composite components are of three types—sequence, selection, or alternation.

There are two types of notations to represent the composite components: a graphical notation which is called a Jackson's Diagram, or Structure Diagram; and a nongraphical notation which is called Structure Text or Schematic Logic (Hic 85, Ja 75, and Ja 83). Jackson's Diagram and structure text notations for sequence, iteration, and selection are given as Figures 10.1 through 10.3.

A sequence has two or more parts which occur once in order. For example, Figure 10.1 shows a Jackson's Diagram and structure text notation for a sequence

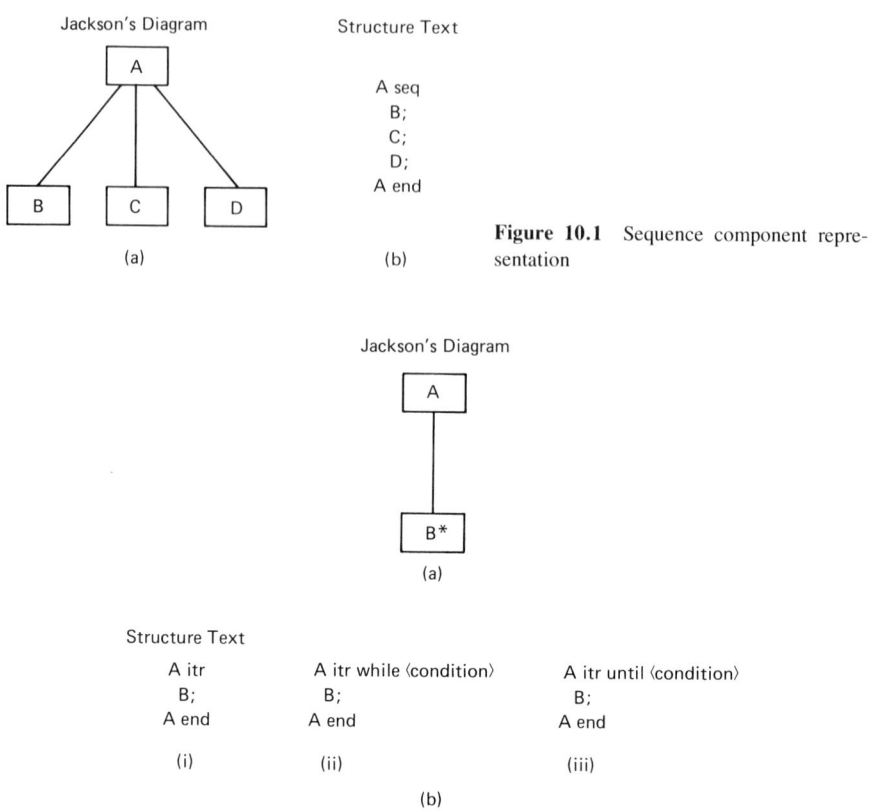

Jackson's Diagram   Structure Text

```
                          A seq
                            B;
                            C;
                            D;
                          A end
```

Figure 10.1 Sequence component representation

(a)   (b)

Jackson's Diagram

(a)

Structure Text

```
   A itr            A itr while ⟨condition⟩      A itr until ⟨condition⟩
     B;                 B;                            B;
   A end             A end                         A end
```

    (i)   (ii)   (iii)

(b)

Figure 10.2   Iteration component representation

component A which consists of one B, followed by one C, followed by one D. In
other words, B, C, and D are components of A.

An iteration component has a part which occurs zero or more times for each
occurrence of the component itself. In Figure 10.2 the asterisk above B inside the
box indicates that component A has an iterated part B; that is, B is repeatedly
executed zero or more times for each occurrence of A. Structure text notation of
component A is given in three different forms as (i)–(iii) in Figure 10.2b.

A selection component has two or more parts, of which one, and only one,
occurs once for each occurrence of the selection component. In Figure 10.3, a
diagrammatic representation and structure text notation of a selection component
A is given. A has parts such as B, C, and D. The small circles in the boxes for B,
C, and D indicate that A is a selection and B, C, and D are its components. It is
also possible to show conditions of selection in the structure text similar to a case
statement as in (ii) and (iii) of Figure 10.3b. A special case of selection is ''null

Jackson's Diagram

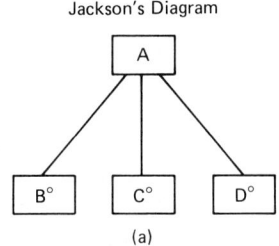

(a)

Structure Text

A sel	A sel ⟨cond-1⟩	A sel ⟨cond-1⟩
B;	B;	B;
A alt	A alt ⟨cond-2⟩	A alt ⟨cond-2⟩
C;	C;	C;
A alt	A alt ⟨cond-3⟩	A alt ⟨else⟩
D;	D;	D;
A end	A end	A end
(i)	(ii)	(iii)

Figure 10.3 Selection component representation

(b)

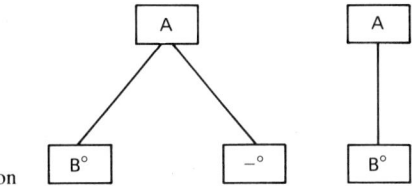

Figure 10.4 Example of null selection

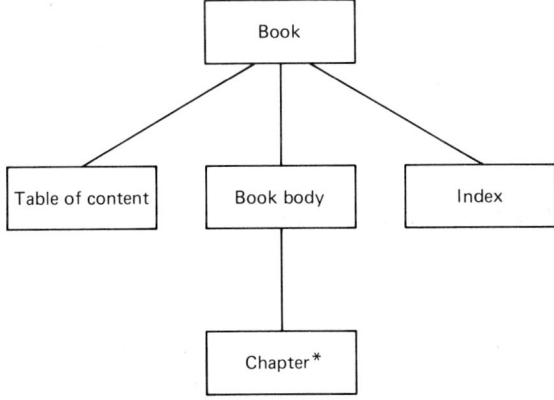

Figure 10.5 Example of iteration in a Jackson's diagram

selection." It means "do nothing" and is represented by $-^0$. Consider the example given in Figure 10.4 in which A has only one selection component, B.

An example of iteration in a Jackson's Diagram is the structure of a textbook in which the table of contents, body of the book, and the index are sequence components and chapter is the iterated part of the body of the book (Figure 10.5).

The composite components and hierarchy are represented in Figure 10.6. In that figure, A is a sequence component, and B, C, and D are its components. C is a selection component; E and F are its components. E and F are iteration components. F is also a sequence component since H and K are its components.

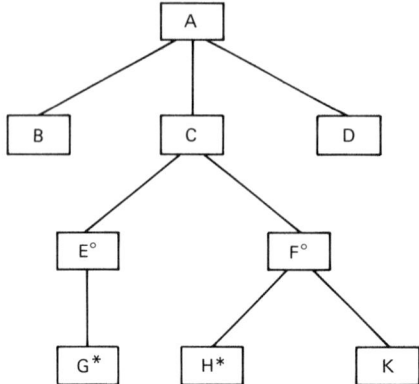

Figure 10.6 Basic structures in Jackson's diagram

A real-life example with Jackson's notation is given in Figure 10.7. In that figure a sales information system is depicted in which a sales clerk is to process sales orders. *Get order record* is an iteration component and it has the following sequence components: *edit customer information, get ordered product information, check ordered quantity,* and *process shippable order. Valid order* and *special order request* are the selection components of *get ordered product information.* Similarly, *process fulfilled order, process partly fulfilled order,* and *process unfulfilled order* are the selection components of *check ordered quantity.* Finally, *process shippable order* is a selection component which has *prepaid order* and *unpaid order* as its elements.

10.3 ENTITY-RELATIONSHIP MODEL (OR CHEN'S E-R MODEL)

E-R Model or Chen's entity-relationship model is a logical model that is used in database analysis and design as well as in the depiction of information systems (e.g., Ch 76, Ch 78, Pan 82). The model has three elements:

- entities
- relationships
- descriptions of entities and relationships or their attributes and values

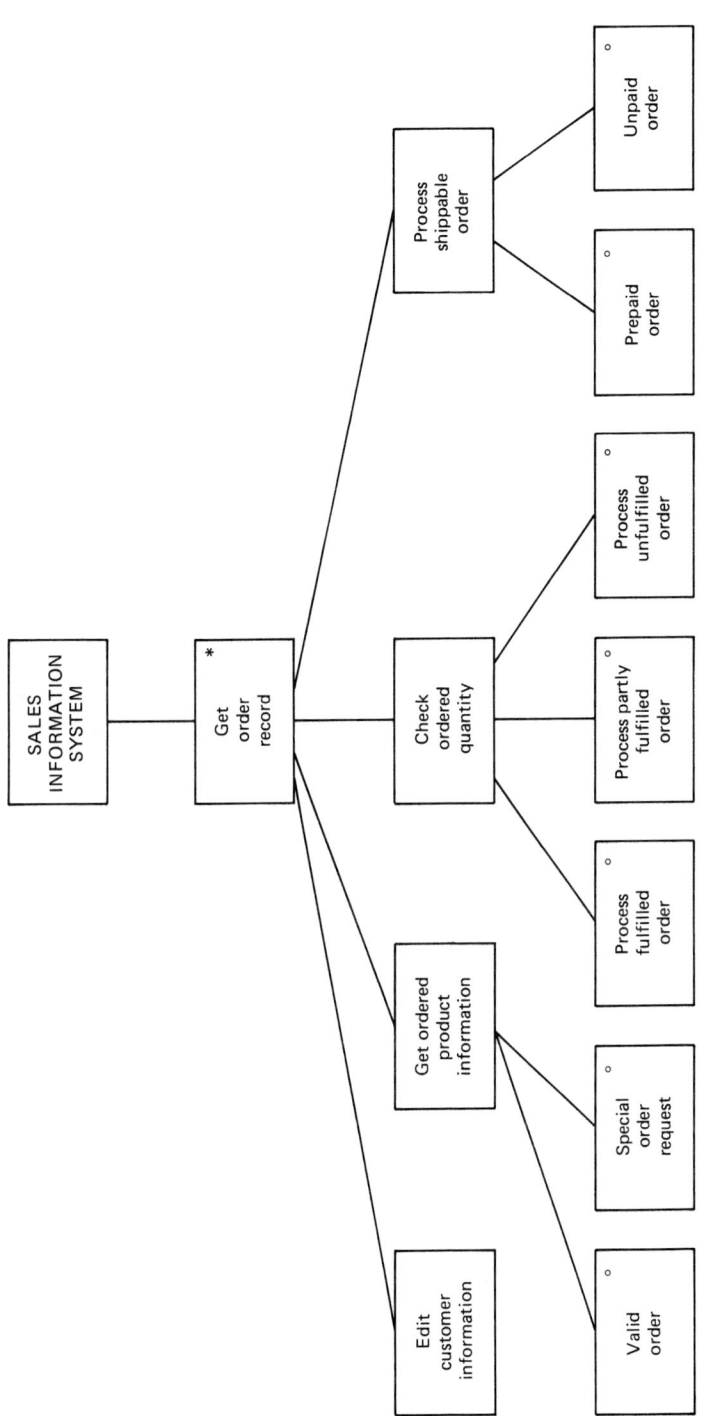

Figure 10.7 A hypothetical system using Jackson's notation

10.3.1 Entities

An entity is a person, place, thing, event, or concept about which information is recorded. For example, in a warehouse the entities are suppliers, parts, shipments, and the like; in a bank, the entities are customers, employees, bank accounts, mortgage loans, and so on. Groups of entities may constitute an entity type, although an entity need not belong to just one type. In an E-R diagram, an entity type is represented by a rectangular-shaped box (some authors prefer an elliptical shape).

10.3.2 Relationships

Relationships—that is, conceptual links—may exist between entities. Relationships are also classified into different types and these relationship types are represented by a diamond-shaped box in an E-R diagram with lines connected to related entity types.

A simple example showing entity and relationship symbols is given in Figure 10.8. In that figure, two rectangles, namely MAN and WOMAN represent entities, and a diamond shape is the relationship, LOVE.

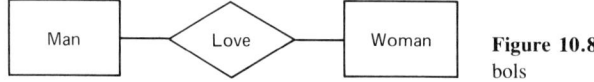

Figure 10.8   Entity and relationship symbols

Relationships in an E-R diagram can be one to one, one to many, and many to many. For example, Figure 10.9 diagrams WORK and PROJ-MANAGER as two different relationship types between two entity types, EMPLOYEE and PROJECT. M, N, and 1 in that figure indicate that there are N projects; M employees

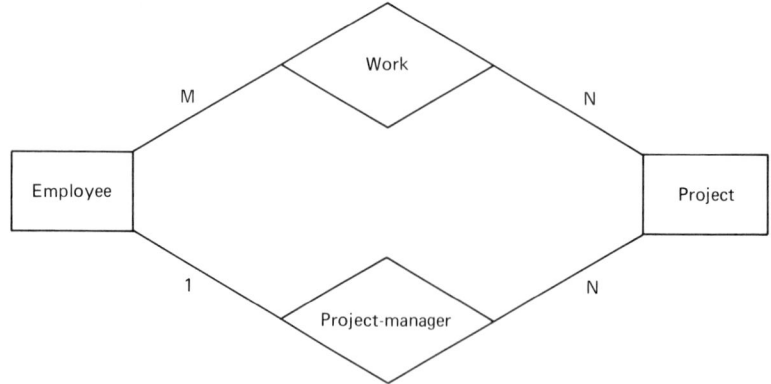

Figure 10.9   Example of entity-relationship (E-R) model

work for these projects, and each project has only one manager. Also an employee can be the manager of many projects. Thus, the PROJ-MANAGER relationship between EMPLOYEE and PROJECT entities is one to many. On the other hand, the WORK relationship between the entities EMPLOYEE and PROJECT is many to many, meaning that each project may consist of several employees and each employee may be associated with more than one project. In the example given in Figure 10.10, the relationship type is MARRIAGE, and it is one-to-one mapping between WOMAN and MAN or among the entity type PERSON.

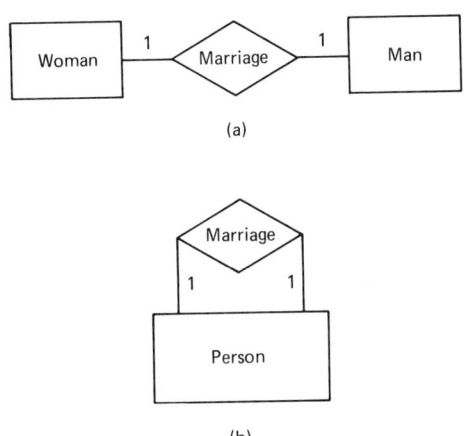

(a)

(b)

Figure 10.10   One-to-one mapping

It is also possible to define a relationship type among more than two entity types. In Figure 10.11, the entities PART, PROJECT, and SUPPLIER are related in a many-to-many relationship: PART-SUPP-PROJ.

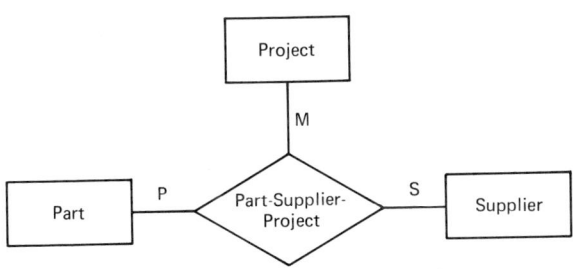

Figure 10.11   Another example of entity-relationship (E-R) model

10.3.3 Descriptions of Entities and Relationships

Every entity has some basic attributes that characterize it. A customer in a bank may be described by such attributes as customer number, name, address, date, and so on. Similarly a house can be described by its size, color, age, and address.

Entities have properties which can be expressed in terms of attribute-value pairs. For example, "SOC-SEC-NO of EMPLOYEE R is 316-88-6972" has SOC-SEC-NO as attribute and 316-88-6972 as the value for the entity EMPLOYEE. Values can be classified into different value types such as SOC-SEC-NO, AGE, COLOR, and QUANTITY. In the E-R notation, a value type is represented by a circle, and an attribute is represented by an arrow directed from the entity type to the desired value type (Figure 10.12). Similar to entities, relationships may have attributes and value types as shown in Figure 10.13.

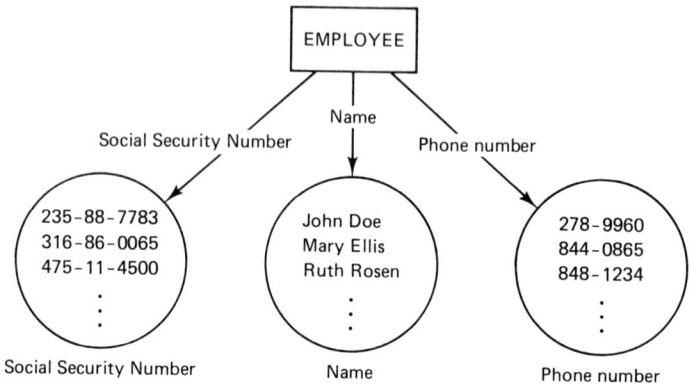

Figure 10.12 Example of entity attributes and value types

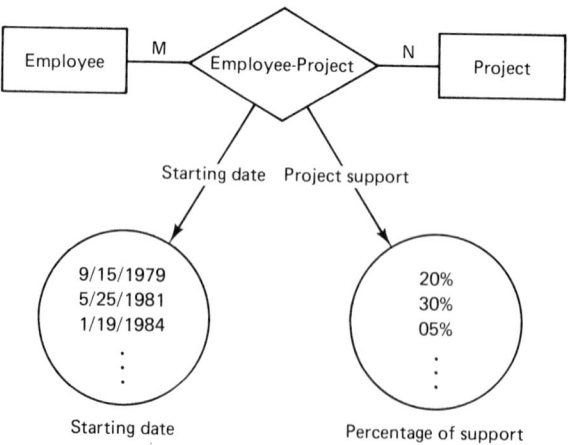

Figure 10.13 Example of relationship attributes and value types

In the example given in Figure 10.13, the STARTING-DATE of an employee in a project depends on both the EMPLOYEE and PROJECT but neither of them alone. Hence STARTING-DATE is an attribute of the relationship EMP-PROJ. Similarly PROJECT-SUPPORT is also an attribute of the EMP-PROJ relationship.

Attribute-value pairs are commonly used to identify entities uniquely. Such attributes of entities are called "entity identifiers" and function the same as primary keys of records in conventional data processing. Relationships are identified by utilizing the identifiers of the entities involved in the relationship.

10.3.4 Special Entity and Relationship Types

Sometimes the existence of an entity may depend on the existence of another entity or entities. Such an entity is called a "weak entity" and it is represented by a double rectangular-shaped box. The relationship box between such entities has "E" to indicate existence-dependent relationship, and an arrow is used to show the direction of dependency. An example is given in Figure 10.14. Figure 10.14 indicates that SPOUSE depends on WORKER. For example, if a worker leaves the company the related spouse data is no longer kept.

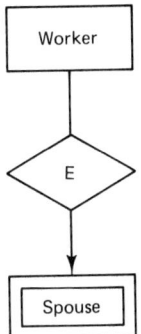

Figure 10.14 Existence dependence example

Another special case occurs if an entity cannot be uniquely identified by its own attribute(s). Its relationships with other entities must then be used for identification. If an entity has such a property, it is then said that it has "ID dependency" on other entities. Such a dependency is indicated by "ID" in a double diamond-shaped box, and the direction of relationship is indicated by an arrow. Entity box is also a double rectangle. An example is given in Figure 10.15.

A town is not uniquely identified unless its state is also defined. As indicated by Chen (Ch 78), most ID dependencies are associated with existence dependencies. However, existence dependency does not imply ID dependency.

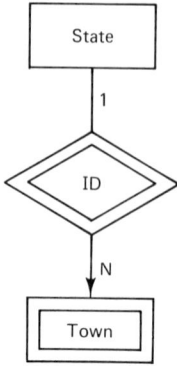

Figure 10.15 ID dependence example

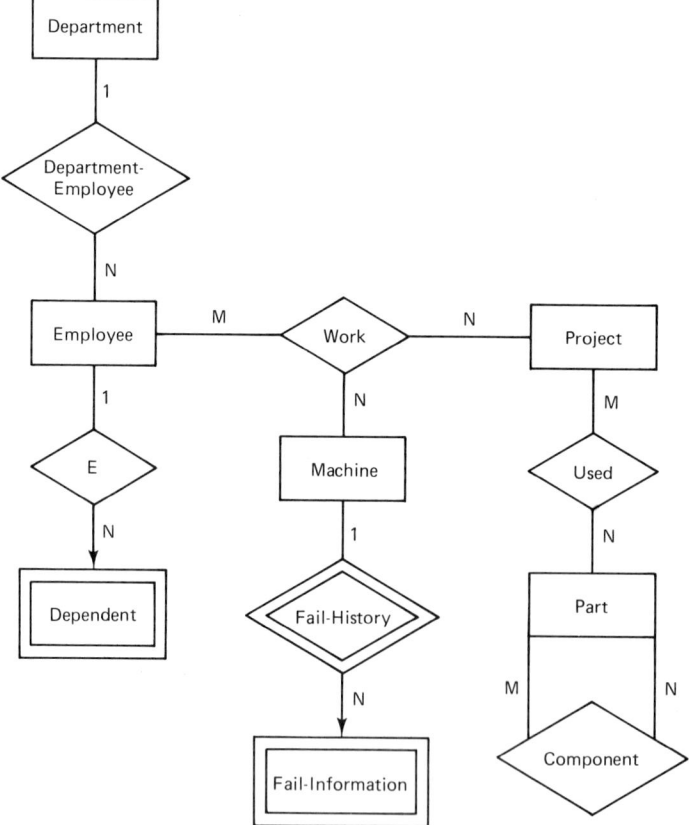

Figure 10.16 Example of an E-R model

Further treatment of special cases in E-R diagrams is presented by Doğaç and Chen (DC 83).

10.3.5 Another Example of the E-R Model

For a further demonstration of the E-R model notation, consider the example given in Figure 10.16 where the entities DEPT, EMP, MACHINE, PROJ, and PART are related through DEPT-EMP, WORK, USED, and COMPONENT relations. DEPENDENT is an existence dependent entity and FAIL-INFO is an ID dependent entity in that example.

10.4 LEIGHTON DIAGRAM

The Leighton diagram is another graphical tool that is available (Pes 80 and Sct 78). Not a very common tool, the Leighton diagram mainly describes the scope

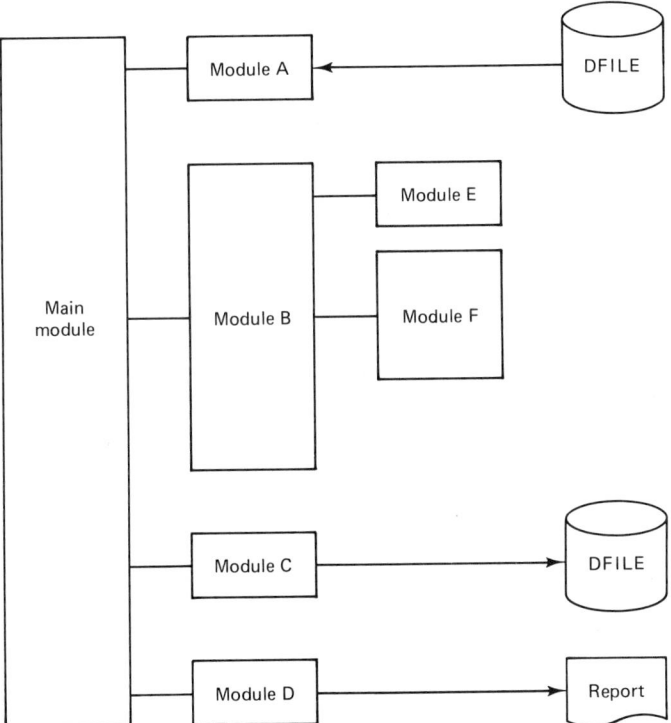

Figure 10.17   Fictitious system outlined by a Leighton diagram

of control, hierarchy, and external interfaces of an information system. Leighton diagrams utilize combinations of

- rectangles
- directed line segment (vectors)
- system flowchart symbols
- connecting lines

A rectangle depicts a module of a system; scope (or span) of control is indicated by the height (i.e., vertical dimension) of the rectangle. Hierarchy is shown by the horizontal dimension of a Leighton diagram, and execution sequence proceeds from top to bottom. A simple example appears in Figure 10.17, which depicts a fictitious system consisting of a main module with four submodules, A, B, C, and D. One may consider Module A as an input module for a disk file. Module B has two process modules, E and F. Module C is an output module for a disk file, and Module D is a reporting module.

SUMMARY

Jackson's notation, Chen's E-R (entity-relationship) model, and the Leighton diagram are other graphical tools that are used in the structured approach in addition to HIPO, DFD, Structured Charts, SADT, and W/O diagrams that were discussed earlier.

Jackson's notation allows us to represent the hierarchy of a system as well as its elementary and composite components. Elementary components are those that are not further decomposed and have no parts. Composite components are of three types—sequence, iteration, and selection. There are two types of notations to represent composite components: a graphical notation, which is called a Jackson Diagram or Structure Diagram, and a nongraphical notation, which is called Structure Text or Schematic Logic.

Chen's E-R model is based on entities, relationships, and their descriptions according to their attributes and values. An entity is a person, place, thing, event, or concept about which information is recorded. A rectangle is used to depict an entity type, which is a group of entities. Relationships—that is, conceptual links—may exist between two or more entities. Relationship types are represented by a diamond-shaped box in an E-R model. Entities and relationships of an E-R model have properties which can be expressed in terms of attribute-value pairs. Existence dependence and ID dependence are special entity types.

The Leighton diagram is another graphical tool that is available. Vertical rectangles, vectors, system flowcharts symbols, and connecting lines are major components of that tool.

EXERCISES

1. What are the basic components of a Jackson Diagram?
2. What are the symbols that are used to represent the basic structures in a Jackson Diagram?
3. What are the basic elements of Chen's E-R model?
4. Define the terms *entity*, *relationship*, *attribute*, and *value*.
5. Investigate the existing student registration system of a college that you are familiar with and represent that system using
   a. a Jackson Diagram
   b. an E-R model
6. Visit a travel agency and examine how the system works. Based on your understanding, represent its major activities using a Jackson Diagram.
7. Give an example for a weak entity and ID dependence.
8. What are the basic components of Leighton Diagrams?

SELECTED REFERENCES

(Ch 76)   Chen, P. P. "The Entity-Relationship Model—Toward a United View of Data," *AMC Transactions on Database Systems*, Vol. 1, No. 1 (March 1976), 9–36.

(Ch 78)   Chen, P. P. *The Entity-Relationship Approach to Logical Data Base Design.* The Q.E.D. Monograph Series, No. 6. Q.E.D. Information Sciences, Inc., Massachusetts, 1978.

(DC 83)   Doğaç, A., and P. P. Chen. "Entity-Relationship Model in the ANSI/SPARC Framework," in *E-R Approach to Information Modeling and Analysis*, ed. P. P. Chen. Elsevier, 1983, 357–74.

(Hic 85) Hiçyilmaz, C. *Jackson System Development Methodology and an Application.* M.S. Thesis, METU-Dept. of Computer Engineering, March 1985.

(Ja 75)   Jackson, M. *Principles of Program Design.* Academic Press, 1975.

(Pan 82) Parkin, A. "Data Analysis and System Design by Entity-Relationship Modelling," *The Computer Journal*, Vol. 25, No. 4 (1982), 401–409.

(Pes 80) Peters, L. J. *Software Design.* Yourdon Press, 1980.

(Sct 78) Scott, L. R. "An Engineering Methodology for Presenting Software Functional Architecture," in *Proc. 3rd. Int. Conf. on Software Engineering.* IEEE Computer Society, 1978, 222–29.

Chapter 11

Nongraphical Tools

So far we have looked at graphical tools that are used for information systems development. Now we will examine some of the nongraphical tools, including the Data Dictionary/Directory (DD/D), Structured English (SE), and pseudocode. As will become clear later, DD is concerned only with *defining* data, whereas SE and pseudocode are concerned with *processing* data.

11.1 DATA DICTIONARY/DIRECTORY (DD/D)

A Data Dictionary essentially specifies the data admissible to a system through naming, classification, representation or structure, usage, and administration of data; it is also referred to as "data about data" or metadata. The range of information which can be held in a Data Dictionary system may be very large. Although the system may be manual, it is usually computerized.

A Data Directory is generally used to specify the location of data within a database and possibly the most appropriate or efficient database path to be followed during access or retrieval. It may also contain descriptions of report formats, screen displays, translation tables, record, file, or schema definitions, and transaction descriptions. In both logic and development, Data Dictionaries precede Data Directories.

Various Data Dictionary/Directory (DD/D) packages are commercially available. Some of these packages are associated with a specific DBMS while others are more general. Some firms develop DD/D software for their own internal use.

11.1.1 Objectives of DD/D

Although a DD/D is a valuable first step in developing a database (DB) and many DBMS (Database Management Systems) include it, it is fast becoming an indispensable tool in information systems analysis and design. In any organization, a given element of data may be defined differently by different users and/or systems analysts. Use of a DD/D system in such an organization improves communication among systems analysts, users, and management in the analysis phase of systems development. The DD/D system clarifies the flow and content of data items through the information system during the design phase and also supports maintenance efforts. Thus a DD/D system is a useful tool throughout information systems development life cycle, as well as being invaluable in documenting the information system.

Kreitzer (Kr 81) introduces a new term, *Information Resource Management* (IRM), which explicitly recognizes information as the lifeblood of today's organization, a resource that should be managed just as any other critical resource. He also states that the Data Dictionary is a central concept of IRM, in effect, the "hub of the wheel" around which revolve the "spokes" of the information resource.

The basic requirements of a DD system are stated as (KL 83):

- Ease in maintenance
- Ease in reporting
- Comprehensive definition and naming conventions
- Adequacy in documentation

Advantages of Data Dictionaries are summarized as control of data, improved system development and control, and automatic generation capability (Ca 78).

Canning (Ca 78) summarized some of the problems with DD systems as follows:

- Perhaps the first problem that confronts the potential user of a DD system might be termed the what-where-when problem. What will it be used for? Where will it be obtained? When will it be installed in the organization?
- In order to fully appreciate the benefits of a DD system, the user organization must have
  - a. a high degree of commitment by management, users, and data processing personnel
  - b. an effective data administration function
  - c. an effective method for planning the introduction of change into information systems

A data management system (DMS) and its relationship with a DD system is elaborated by Canning (Ca 81). A DMS is defined as a system that includes (1) a DD for defining new files, records, and fields, plus indexes for accessing the records, plus a means for allocating disk space for the files; (2) a means for creating screen formats for inputting and validating data; (3) a means for entering data and updating the database; (4) a record selection and sorting capability; (5) a query capability; (6) a report formatting and column totalling capability; and (7) a means where by specific application logic can be expressed. Thus a DMS can perform most of the routine aspects of a data processing application and in so doing relies heavily on its data dictionary facility.

11.1.2 Content of DD

A data item in a DD may be an element or a group. An element is the lowest level data such that it cannot be decomposed into smaller pieces. Elements may sometimes be components of a group. A group definition shows the component data elements that make up the group and the relationships among them. A set of relational operators are used to define the composition of a group data element, as will be discussed in the next subsection.

The information that is kept in a DD for each data element may include the following identifiers: naming, classification, representation, usage, and administration (e.g., BCS 77).

Naming the information of an element is basically listing its aliases, that is, the variety of names that may be used at different times and in different places to identify the element.

Classification information includes the description of the item in natural language, its ownership, item type (if it is a group[1] or elementary item), privacy and security considerations, and authorization definitions.

Representation information or format may include item length, picture,[2] its composition if it is a group item, and processing units.

Usage information describes the use of the element quantitatively, giving its range of values, frequency of use, conditional values, if any, and so on.

Finally, administration information includes the resources used or required by the element and the processing mode, that is, batch, time sharing, and transaction processing.

11.1.3 Composition Definition of Group Data Items in DD

As noted previously, elements of group data items of DD and their interrelations are defined by means of some formulas using relational operators. The commonly used operators are summarized in Table 11.1 (see DeM 78 and Pes 80).

---

[1]A data item is called a group item if it consists of more than one elementary item.

[2]In COBOL, an elementary item is described by a picture clause, which is a description of the number and type of characters that make up the item.

**TABLE 11.1   Relational Operators for Composition
Definitions**

Symbol	Operation
=	IS COMPOSED OF (or IS EQUIVALENT TO)
+	AND
[ ]	EITHER-OR (i.e., selection)
{ }	ITERATIONS OF
( )	OPTIONAL

In addition to these symbols, a pair of asterisks (*) is used for comments and a pair of quotes for nonnumeric literals which are constants not being used for arithmetic operations. A vertical bar (|) is used for separating options. The upper and lower limits of iterations are also shown outside the iteration braces. For example, iterations from 1 to 10 may be shown by

$$\underset{1}{\overset{10}{}}\{DATA\ ELEMENT\} \quad \text{or} \quad 1\{DATA\ ELEMENT\}10$$

Default limits are 0 and ∞, meaning that it will be repeated zero or an undefined number of times.

Examples of composition definition using relational operators are as follows for the data items CLASS LIST and PAYMENT:

CLASS LIST = {NAME}

NAME = (INITIAL) + FIRST NAME + LAST NAME

FIRST NAME = 1{ALPHABETIC CHARACTER}10

LAST NAME = 1{ALPHABETIC CHARACTER}30

$$PAYMENT = \begin{bmatrix} \text{``CASH''} \\ \text{``PERSONAL CHECK''} *WITH\ APPROVAL* \\ \text{``CREDIT CARD''} \end{bmatrix}$$

or

PAYMENT = [``CASH''|``PERSONAL CHECK''|``CREDIT CARD'']

As shown above CLASS LIST consists of NAME. Each NAME has an optional INITIAL, a FIRST NAME and a LAST NAME. FIRST NAME has one to ten letters, and LAST NAME has one to 30 letters. PAYMENT can be in terms of

```
Name         :   LAST-NAME
Aliases      :   None
Description  :   It is the last name of a student
                 enrolled in a course
Format       :   It consists of alphabetic characters
Composition  :   1 {A | B | C | · · · | X | Y | Z} 30
Location     :   NAME field
Control Info :   None
Other        :
```

Figure 11.1 DD example

CASH or PERSONAL CHECK or CREDIT CARD. An example of DD is given in Figure 11.1.

11.1.4 Commercially Available Data Dictionary/Directories

DD/Ds are frequently used in combination with DBMSs. As stated earlier, this is not necessary and there are numerous DD/D users who still have no DBMS. Also there are a number of independent DD/D systems developed by specialized software houses. Some of the commercially available DD/D systems include Control 2000 of MRI Systems, Data Catalogue of Synergetics, Data Dictionary of Cincom, Datamanager of MSP Inc., DB/DC Dictionary of IBM, Dictionary 204 of Computer Corporation of America, IDMS Dictionary of Cullinane, Lexicon of Arthur Anderson, and UCC TEN of University Computing Co. (Scl 77, Se 80).

11.2 STRUCTURED ENGLISH

Structured English (SE) is a very limited, highly restricted subset of the natural language English. In a way SE resembles a programming language and it is an efficient tool to describe an algorithm. It is quite similar to pseudocode that will be discussed next. Because of this similarity SE and pseudocode are often mixed. SE is a better tool to express an algorithm if the main concern is *user* communication. Pseudocode, however, is a better tool if the concern is *programmer* communication. Neither tool is efficient, however, if the algorithm to be expressed has many decisions. A flow chart, decision table, or decision tree may be a better tool in such a case. SE's intermediary position is shown illustratively in Figure 11.2 (Yo 81).

Although variations abound and there is not yet a standard SE, the major characteristics of this tool may be summarized as:

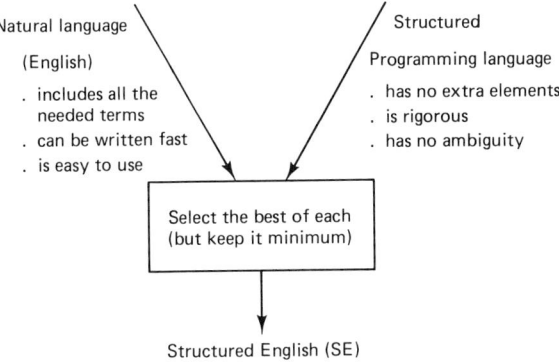

Figure 11.2   A definition of structured English

Limited format for expression.   Simple imperative sentences and/or algebraic expressions are used. Examples of simple imperative sentences are:

Read Master-File.

Multiply Hrs by Wage to get Gross-Pay.

Limited volume of vocabulary.   Sentence objects must be defined in the DD (Data Dictionary), and certain reserved words should be used for logic formulation. For naming the sentence object, a COBOL-like notation may be used.

Limited number of building blocks.   Sequence, selection, and iteration are the basic structures used to put sentences into blocks.

Sequence.   Sequence structure is a sequential collection of imperative sentences, as the following series indicates:

Read First-Record.

Initialize District-Fields.

Add 1 to Counter.

For long algorithms, grouping of some statements into a named block and treatment of such a block as a single statement may be practical. For example, define all individual statements needed to prepare total sales as a block named Total-Sales and reference this block by a single statement:

Perform Total-Sales.

Selection.   For selection (or decision) logic, an if-then-else structure is used, and indentation makes its logic clearer. An example of logic that updates a master file may be expressed as:

```
if Trans-Type = "CHANGE"

    then modify New-Record
    get  Next-Transaction
    else if Trans-Type = "DELETE"

             then delete New-Record

                  get Next-Transaction

             else print Error-Msg

                  get Next-Transaction.
```

The general form of this structure may be expressed as

```
if <condition>
    then block-1
    else block-2.
```

Iteration.   Iteration (or repetition) logic defines a block (or group of SE statements) that is executed repeatedly until a termination condition is satisfied.

There are two general forms for expressing iteration logic:

1. For each <item>
      block-1.

2. Perform until <condition>
      block-2.

The following examples demonstrate these forms:

1. For each Record
      Process-Record

2. Perform until End-of-File
      Process-Record
      get Next-Record

Process-Record in these examples is a block consisting of operations defined elsewhere in the SE statements.

SE is flexible in notation and may consist of various styles. Some of these styles, in addition to the "common style" given above are summarized as follows. The process is group totals of branch office sales for a company.

Common style

```
Print Corporate-Heading
Initialize Corporate-Total
For each Branch
        Print Branch-Heading
        Calculate Yearly-Sales
        Find Max-Sale
            Max-Sale-Val = Sales-Val (1)
            For each Month
                if Sales-Val(Month-No) > Max-Sale-Val
                    then Max-Sale-Val = Sales-Val(Month-No)
        Print Branch-Results
        Add Yearly-Sales to Corporate-Total
Print Corporate-Total
```

Here it is assumed that monthly sales values for branch offices are available. Maximum monthly sales value is computed in a loop checking the monthly sales. The yearly sales total and the maximum monthly sales value for each branch office are computed and printed in an outer loop.

Code style (capitalized common style)

```
PRINT CORPORATE-HEADING
INITIALIZE CORPORATE-TOTAL
FOR EACH BRANCH
        PRINT BRANCH-HEADING
        CALCULATE YEARLY-SALES
            .
            .
            .
```

Outline style (numbered common style)

```
1. Print Corporate-Heading

2. Initialize Corporate-Total

3. For each Branch
   3.1 Print Branch-Heading
   3.2 Calculate Yearly-Sales

   3.3 Find Max-Sale

       3.3.1 Max-Sale-Val = Sales-Val (1)

       3.3.2 For each Month

                 if Sales-Val (Month-No) > Max-Sale-Val

                 then Max-Sale-Val = Sales-Val (Month-No)

   3.4 Print Branch-Results

   3.5 Add Yearly-Sales to Corporate-Total

4. Print Corporate-Total
```

Narrative style

First print Branch-Heading and then initialize Corporate-Total value.

Next, considering each Branch separately, do the following operations:

Calculate Yearly-Sales, find Max-Sale, print Branch-Results, Add Yearly-Sales to Corporate-Total. At the end, print Corporate-Heading and Corporate-Total.

11.3 PSEUDOCODE

Pseudocode is an alternative to Structured English and it is similar to such programming codes as COBOL, PL/1, FORTRAN, or Pascal. It is therefore easy for programmers to use and to understand but is not suitable for nonprogrammers.

When SE is used, some details such as opening or closing files, initializing counters, or setting flags are usually not included. With pseudocode they are included. However, the pseudocode user is not concerned with a number of language-dependent details such as the difference between real and integer numbers in FORTRAN or DCL statements in PL/1 or the distinction between subscripts and indexes for table manipulation in COBOL, that is, the definition of data in any language.

Like SE, there is not a standard, universal pseudocode; various versions exist. In any pseudocode version, however, the three basic structures, namely sequence, selection, and iteration, are often included.

Sequence.   Sequence is a collection of various statements. Input/Output instructions are explicitly defined in pseudocode; that is, such statements as

Read data from source

and

Write data to destination

can be specified. It is also possible to group and name a number of pseudocode statements and treat them as a single block using the perform verb.

Selection.   The general form of a selection or decision pseudocode block may be written as

```
If <condition>
    then perform block-1
    else perform block-2
Endif
```

Thus a selection block begins with an *If* and ends with *Endif.*

The CASE structure is a common tool that is used if a problem involves a selection from among several alternative paths. Its general form is:

```
Selected variable
          CASE (value-1)   block-1
          CASE (value-2)   block-2
               .
               .
               .
          DEFAULT CASE      block-n
Endselect
```

Iteration.   In pseudocode, one is more concerned with the various forms of repetitive logic than SE. The basic idea of repetitive logic is that a block is executed repeatedly until a termination condition is satisfied. There are three forms for repetition logic in pseudocode:

Do while structure: This is the commonly used structure for iteration. Its general form is

```
While <condition> do
          perform block
Endwhile
```

As is evident, While and Endwhile delimit the block.

Repeat until structure: Another structure for repetition logic is known as the repeat until structure. Its general form is

```
Repeat
          Perform block
Until   <condition>.
```

Do structure: Sometimes repetition logic in pseudocode is expressed similar to FORTRAN and PL/I. The general form is

```
Do index = initial to limit
        Perform block
Enddo
```

Again, please note the indentation and placement of *Enddo* statements. Statements within a block are indented and *Enddo* aligns with *Do*.

SUMMARY

The most commonly used nongraphical tools for information systems development are the Data Dictionary/Directory (DD/D), Structured English (SE), and pseudocode.

A data dictionary is a collection of information about naming, classification, representation or structure, usage, and administration of data. It is also referred to as "data about data" or metadata. Various DD/D software packages are commercially available. Use of such a system in the analysis phase improves communication among systems analysts, users, and management. A DD/D system clarifies the flow and content of data items through the information system during the design phase. A data item in a DD may be an element or a group. A set of relational operators is used to define the composition of group data.

Pseudocode is an alternative for SE. Pseudocode is similar to a programming code and therefore it is a preferable communication tool between systems analysts and programmers. Like SE, pseudocode has no universal standard; various versions exist. In any pseudocode version, however, the three basic structures—namely sequence, selection, and iteration—are often included.

EXERCISES

1. What are nongraphical tools in information systems development?
2. What is a DD/D?
3. What are the objectives of DD/D during an information system development process?
4. What is an element data item?
5. How would you define the composition of a group data item?
6. State any three of the relational operators that are used in DD composition definition.
7. What is the meaning of the following composition definition? CUSTOMER-NAME = 1{(TITLE) + FIRST-NAME + LAST-NAME} 50
8. Prepare a DD description for an employee record which consists of ID No., Social Security No., Name, Department, Position, and Starting Date.
9. Using Structured English first and then pseudocode, describe a module to calculate state sales tax charges for a customer's total purchases.

SELECTED REFERENCES

(BCS 77)   The British Computer Society. *Data Dictionary Systems Working Party Report.* DDSWPR, March 1977.

(Ca 78)   Canning, R. G. "Installing a Data Dictionary," *EDP Analyzer*, Vol. 16, No. 1 (January 1978).

(Ca 81)   Canning, R. G. "A New View of Data Dictionaries," *EDP Analyzer*, Vol. 19, No. 7 (July 1981).

(DeM 78)   DeMarco, T. *Structured Analysis and Systems Specification.* Yourdon Press, 1978.

(KL 83)   Kahn, B. K., and E. W. Lumsden. "A User-Oriented Framework for Data Dictionary Systems," *DataBase*, Fall 1983, 28–36.

(Kr 81)   Kreitzer, L. W. "Data Dictionaries—The Heart of IRM," *Infosystems*, No. 2 (1981), 64–66.

(Pes 80)   Peters, L. J. *Software Design*. Yourdon Press, 1981.

(Scl 77)   Schussel, G. "The Role of the Data Dictionary," *Datamation*, Vol. 23, No. 6 (June 1977), 129–42.

(Se 80)   Semprevivo, P. *Using Data Dictionary/Directories*. Auerbach Publishers Inc., Portfolio No: 32-04-09, 1980.

(Yo 81)   Yourdon, Inc. *Structured Analysis/Design Workshop*, Edition 7.2, July 1981.

Chapter 12

Evaluation and Refinement of Qualities of a System Design

12.1 GENERAL

As noted before in early chapters, the major concern of a structured system is that it is "modular." In other words, it should be partitioned into separately named and addressable elements or components that are integrated to meet the system requirements. Each of these elements is called a module and, depending on their level in the system, they may be defined as subordinate or superordinate modules. Analogous to the structure of business or military organizations, a modular system should also have a "hierarchy" to identify the top-level modules in charge of controlling activities and related decision-making processes and the low-level modules in charge of detailed operations. "Module independence" is another important property in addition to modularity and hierarchy of a structured information system. System development, testing, and maintenance are simplified if the modules of a system are independent or relatively so. There is a close relationship between module independence on the one hand and modularity and hierarchy on the other; in fact, one may state that module independence is a critical factor in an information system design activity.

The modularity concept in systems design is a result of the classical "divide and conquer" logic. Yet, one also feels that after a certain number of modules—that is, the number of divisions—the integration of the modules of the system becomes more difficult. Hence there should be an optimum number of modules for any system being designed. A system with a relatively small number of modules is termed "undermodular," and a system with an excessive number of modules is called "over-modular." Unfortunately, we have no formula to determine that optimum number and we are forced to use our own intuition or experience. Even if one had that number, it is not enough to finalize a design. Assume a system with five modules, m_1, m_2, . . . , m_5, to be designed. Consider three design alternatives A_1, A_2, A_3 as shown in Figure 12.1. The figure illustrates only three of the many possible configuration alternatives using five modules. A system designer must pick only one of them. Obviously then, not only the number of modules in the sytem, but also their communications and hierarchical relations should be considered in finalizing an information system design. Fortunately, there are some qualitative tools to be used as guidelines to evaluate and refine the properties of an information system design and to compare the overall system with alternatives.

Two such tools, *coupling* and *cohesion*, are used to measure module independence within the system. Considering the vital function of module independence in an information system design—and hence in information system development—one may realize the significance of coupling and cohesion. Coupling is a measure of the relative interdependence among two or more modules. Cohesion (or strength or binding) measures the degree that each module of an information system performs a single, problem-related function. In other words, it indicates the relative functional strength of a module in a system. The function of a module is a description of the input to output transformation that occurs when the module is called. The function is related not only to the operations performed in that module, but also to the functions of any modules called by that module.

Although modularity, hierarchy, module independence, and coupling-cohesion measures are the key points in information system design activities, there

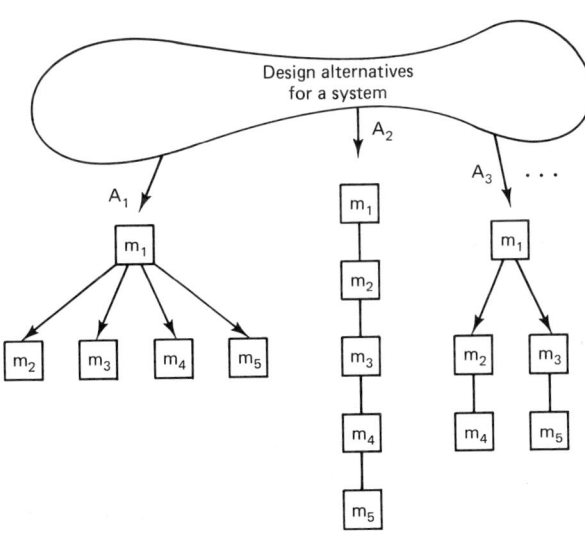

Figure 12.1 Design alternatives for a system

are some additional guidelines that one can use to evaluate and refine the qualities of an information system design. Some of these guidelines are given here:

- Factoring and module size
- Information hiding
- Decision splitting
- System shape or system morphology
- Error reporting
- Editing
- Restrictivity/generality
- Fan-in/fan-out

The following are used for the evaluation of

a. an individual module

- its size
- fan-in/fan-out
- its cohesion level
- its information-hiding property
- its restrictivity/generality

b. a group of modules

- coupling
- factoring
- decision splitting
- error reporting
- editing

The systems design is refined by

- factoring
- combining
- rearranging

As we shall see in the following chapters, various methodologies are available for information systems development. Because these methodologies are based on different approaches, the resulting system design alternatives will be different for the same set of information systems requirements. The guidelines we have listed are used to evaluate and refine the qualities of the resulting information systems design alternatives. In the following sections we discuss these guidelines in greater detail before we go on to individual systems development methodologies. Remember that the design phase is really the crux of the whole information system development efforts.

12.2 COUPLING

Quite often coupling and cohesion concepts are presented by means of structure charts. Clearly, the use of data couples and control couples will simplify the explanation of these concepts. It is, however, incorrect to consider coupling and cohesion only in relation to structure charts. They are general concepts that are applicable to any modular system.

As noted earlier, coupling is a measure of the degree of interdependence among the modules of a system. Modularization of a system will minimize coupling and in turn facilitate understanding of how individual modules function and lessen the "ripple effect" in the system during maintenance. If some modules are highly coupled, one needs an overview of the functioning of all, even if only one is needed for maintenance purposes; and, of course, a change in one module is bound to have some effects on modules coupled to it. One may, therefore, conclude that a minimum degree of coupling between the modules of a system will result in a well-designed, better functioning system.

Data coupling, stamp coupling, control coupling, common coupling, and content coupling are coupling types. Any two modules of a system have *data coupling* if necessary data is communicated between them. The data may be a single data item or an element of an array. Data that is communicated between modules without any use is called "tramp data" and should be avoided.

Two modules of a system are said to be *stamp coupled* if they communicate a group of related data items such as a record consisting of various fields or an array consisting of elements. Obviously any change either in format or structure of a field in a record or an element in an array will have an effect on the modules that use that record and/or array. Not all but only some fields of a record or some elements of an array may be needed in a module. Still, the whole record and/or array will be communicated. Thus stamp coupling tends to expose a module to more data items than it needs. This may not always be desirable because of the high probability of error and the inefficiency of the system. Another potential problem for stamp coupling is "bundling." Bundling involves collecting fairly unrelated data and/or control into an artificial data structure. Because bundling increases the difficulty of understanding the functioning of modules, it should not be used in a structured system.

Two modules of a system are *control coupled* if one of the modules communicates a piece of information, usually a flag, intended to control the internal logic of the other module. Two types of flags are the descriptive flag and the control flag. A descriptive flag has an adjective (e.g., "code is alphabetic" or "new customer data is at the end"). A control flag, on the other hand, has a verb such as "initialize counters," or "delete customer record."

Two or more modules of a modular system are *common coupled* if they share data that is held in a common area. Here the word *common* is used in the same sense as COMMON in FORTRAN or as global variables in PL/I, and similar to the way the REDEFINES clause is used in Data Division of COBOL. Such a coupling obscures the functioning of modules and, therefore makes maintenance more difficult. Also, an error or change in a module using the common data area may be transmitted to other modules, and thus common coupling should be avoided whenever possible.

The worst type of coupling between any two modules is called *content coupling*. To express its undesirable properties, such coupling is sometimes referred to as "pathological or sick coupling." Such a coupling will occur if a module refers to or modifies data contained inside another module without communicating with that module, if it alters a statement in another module, or if it branches or falls through into another.

If two modules have more than one type of coupling, it is the lower level type that will affect the system. In practice, it is not needed to determine the precise level of coupling. Rather, it is important to minimize the coupling so that module independence is increased.

The coupling types, their relative levels, degrees, and qualities are summarized in Table 12.1.

The table indicates that any two modules of a modular system may have (1) no direct coupling or (2) any coupling ranging from data coupling to content coupling. It is impossible to have a system in which all modules have no coupling. The objective is to minimize the coupling level of modules. Level values in Table 12.1 are given only to indicate relative values; they should not be used for any numerical operations.

TABLE 12.1 Coupling Types

Coupling Type	Level	Degree	Quality
no direct coupling	0	low	best
data coupling	1	↑	↑
stamp coupling	2		
control coupling	3		
common coupling	4		
content coupling (or pathological coupling)	5	high	worst

12.3 COHESION

Cohesion, module strength, and binding are all terms signifying one method of evaluating a system design. As defined earlier, cohesion is a measure of the degree by which each module carries out a single, problem-related, and well-understood function. If a module reaches that objective, it is defined as "conceptually whole" or as "functionally cohesive." Ideally we would like to have functionally cohesive modules. There are modules, however, that perform not a single function but a group of functions. Such modules are defined as "fragmented modules." Depending on the degree of fragmentation, that is, on the strength of association of elements within a module, one may define different types of cohesion for such modules.

There is a close relationship between coupling (that is, what goes on between modules) and cohesion (that is, what goes on inside individual modules). A system with highly cohesive modules will have low coupling, a much desired objective in systems design.

Stevens-Myers-Constantine (SMC 74) and Yourdon (Yo 81) have developed a scale of cohesion given as Table 12.2 which also includes a maintainability range.

In the literature, functional cohesion is also referred to as "external cohesion"; the remaining six cohesion types are grouped together as "internal cohesion." The six levels of internal cohesion may be further grouped as:

Data-Oriented Cohesion Types

- Sequential
- Communicational

A module has *sequential cohesion* if its elements are involved in a sequence of activities such that result or output of an activity is an input to the next activity of that module. Such a module is easily maintainable.

A module is said to have *communicational cohesion* if its elements contribute to activities that use the same data item as an input. Such a module is quite maintainable.

Sequential and communicational cohesion types are quite similar. The main difference between them is that order of execution is unimportant for communi-

TABLE 12.2   Scale of Cohesion

Cohesion Type			Level of Cohesion	Ordinal Values	Maintainability
External Cohesion	Functional		1	10	Best
Internal Cohesion	Sequential	Data	2	9	
	Communicational	Oriented	3	7	
	Procedural	Time	4	5	
	Temporal	Oriented	5	3	
	Logical	Suite	6	1	
	Coincidental	Oriented	7	0	Worst

cationally cohesive modules, whereas the sequential type demands a certain pattern of execution.

Time-oriented cohesion types

- Procedural
- Temporal

Modules which have *procedural cohesion* and *temporal cohesion* are called time-oriented cohesive modules. In a procedurally cohesive module, control flows from one activity to the next, and activities within the module may be different and have little relationship among them.

A module has temporal cohesion if its elements are involved in activities of different groups that are related only in time, such as an initialization module in which rewind tape, set counter, clear table, and set switch operations are performed. Similar to the difference between sequential and communicational cohesion, the order of execution of activities is more important in procedurally cohesive modules than in temporally cohesive modules.

Suite-oriented cohesion types

- Logical
- Coincidental

Modules with *logical* and *coincidental cohesion* are said to have suite-oriented cohesion. Such modules are highly dependent on other modules for their activities and decisions.

A module has logical cohesion if its elements are not related by flow of data or by flow of control, but related only to tasks of the same general category. Finally, a module has coincidental cohesion if its elements have no meaningful relation at all.

Quite often it is difficult to determine the exact level of cohesion of a module. However, this is unnecessary since the main objective is to design systems with highly cohesive modules.

12.4 ADDITIONAL DESIGN GUIDELINES

As stated earlier, we need some further guidelines in addition to coupling and cohesion to design better systems. In the following these guidelines are briefly discussed (e.g., Jo 80, Pes 81, Pr 82, and YC 79):

Factoring and module size.   Factoring is also known as decomposition, partitioning, or explosion. It is the separation of a function in one module into a new module of its own. The objectives in factoring are to reduce module size, to simplify development and implementation of individual modules, and to improve the coupling/cohesion property of modules.

Information hiding.   The principle of information hiding proposed by Parnas (Pas 72) suggests that modules should be specified and designed so that information within a module is inaccessible to other modules that have no need for such information. The practice provides benefits during the testing and maintenance of individual modules.

Decision splitting.   In general, recognition of what action to take and execution of that action are the two parts of a decision. A decision split occurs if the two parts of a decision are separated and placed into two different modules. Decision splits should be avoided as much as possible.

System shape or system morphology.   Major features of system shape are depth, width, and overall morphology (Figure 12.2). Depth of a system is defined as the number of levels in the hierarchy. Width is the maximum number of modules at any level of the system under consideration. For the overall morphology of a system, Yourdon and Constantine report that most well-designed systems have a shape that could be likened to a cigar, a flying saucer, or a mosque (YC 79, p. 157).

Error reporting.   As a general rule, errors should be reported by the module which detects the error and determines its type.

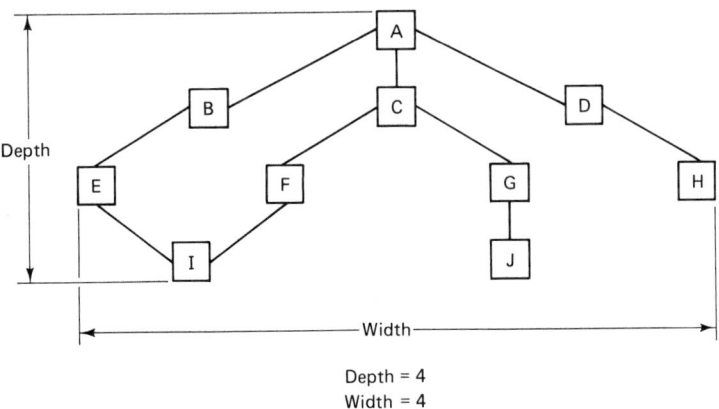

Depth = 4
Width = 4

Figure 12.2   System shape

Editing. Editing of a piece of input data should be done in the following order (Jo 80):

- Edit known before unknown. If a person entering a field realizes that a mistake has been made, give the person a chance either to correct it or to cancel the entry.
- Edit syntactic before semantic. Check the data's format before checking its sense. For example, UT + H would be syntactically incorrect, whereas UT × H would be syntactially correct, but semantically incorrect for UTAH.
- Perform single editing before cross-editing. Cross-validate only those fields that are individually correct.
- Edit internal before external. Make sure that you verify the syntax and semantics of all the fields of a given record before trying to use that record in something else.

Restrictivity/generality. A module of a system should not be neither too restrictive nor too general. As stated in Jones (Jo 80), a restrictive module has one or more of the following characteristics:

- It performs a needlessly specific job.
- It deals with restrictive data values, types, or structures.
- It makes assumptions about where or how it's being used.

An overgeneral module has one or more of the following characteristics:

- It performs a very broad job.
- It deals with too many data types, values, or structures.
- It reads in, or takes as a parameter, data that is unlikely to change.

Fan-in/fan-out. Fan-in of a module is the number of modules that call it. Fan-out or span of control, is the number of direct subordinates to a module. As a general rule, fan-in of modules should be as high as possible (i.e., reusable modules) and fan-out of modules should not be more than 7 ± 2. Figure 12.3 has an example of fan-in and fan-out.

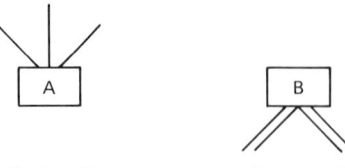

Fan-in = 3   Fan-out = 4   **Figure 12.3** Fan-in/fan-out of modules

12.5 EXAMPLES

Consider the system described by the structure chart given in Figure 12.4. The modules of the system and their functions are as follows:

MAIN : sends a control couple (FLAG-1) to INPUT-DATA module and another control couple (FLAG-2) to EDIT-DATA module; gets edited input record (ED-REC), passes it to PROCESS module; gets data items RESULTS from PROCESS module and passes it to FORMAT module; it receives formatted RESULTS data items, FMT-RESULTS, from FORMAT module;

INPUT-DATA : gets input record (IN-REC), passes it to the editing module (EDIT-DATA);

PROCESS : gets edited input record (ED-REC), processes it, and yields RESULTS data items;

FORMAT : gets RESULTS data items and puts them into a proper format to yield formatted results (FM-RESULTS);

GET-DATA : sends input record (IN-REC) to the calling module (INPUT-DATA);

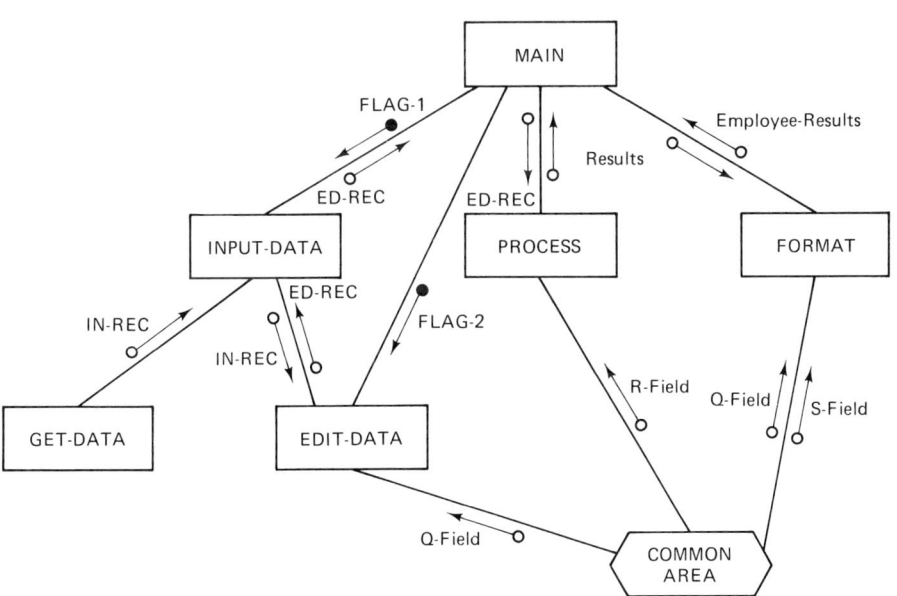

Figure 12.4 A sample system

EDIT-DATA   :   gets input record (IN-REC), edits it, and passes the
                       edited record (ED-REC) back to the calling module (IN-
                       PUT-DATA).

The coupling characteristics of the modules of the system described in Figure 12.4
may be summarized in tabular form as Table 12.3.

TABLE 12.3   Coupling Characteristics of the Sample System

Module	Module	Coupling Type	Communicated Item(s)
MAIN	INPUT-DATA	control coupling	control couple (FLAG-1)
MAIN	PROCESS	stamp coupling	record
MAIN	FORMAT	data coupling	data item (field)
MAIN	GET-DATA	—	no direct coupling
MAIN	EDIT-DATA	content coupling	control transfer into another module
INPUT-DATA	GET-DATA	stamp coupling	record
INPUT-DATA	EDIT-DATA	stamp coupling	record
INPUT-DATA	PROCESS	—	no direct coupling
INPUT-DATA	FORMAT	—	no direct coupling
PROCESS	GET-DATA	—	no direct coupling
PROCESS	EDIT-DATA	common coupling	common data items
PROCESS	FORMAT	common coupling	common data items
FORMAT	GET-DATA	—	no direct coupling
FORMAT	EDIT-DATA	common coupling	common data items

TABLE 12.4   Cohesion Types of Various Sample Modules

Sample Module Name	Function	Cohesion Type
A	gets transactions, edits them, returns them	sequential
B	using employee number, it determines employee name, determines employee address, determines employee position, and returns name, address, and position	communicational
C	performs initialization, reads data file, edits, writes, and closes the file	procedural
D	daily activities such as getting up, having breakfast, driving to the office, doing the daily chores, and returning home	temporal
E	drink tea, drink coffee, drink coke, drink water	logical
F	study, go to concert, visit friends and relatives, paint car, read paper, play tennis, and eat lunch	coincidental

Recall that module cohesion (or strength) can be classified as data-oriented cohesion (sequential or communicational cohesion), time-oriented cohesion (procedural or temporal cohesion), and suite-oriented cohesion (logical or coincidental cohesion). Some sample modules, named A thru F, are tabulated as Table 12.4 to describe their functions and specify their cohesion types accordingly.

SUMMARY

Modularity and hierarchy are two major characteristics of a structured system. In addition to them, module independence is another property that a structured system must have. In finalizing an information system design, the optimum number of modules, their communications, and hierarchical relations with the system should be considered. There are some qualitative tools available to evaluate and refine the properties of an information system design and to compare it with other system alternatives. Coupling and cohesion are two such tools.

Coupling is a measure of the degree of interdependence among the modules of a system. It should be minimized in order to facilitate understanding how individual modules function and to minimize the "ripple effect" during system maintenance. There are various coupling types; data coupling is the most desired and content coupling (or pathological coupling) is the least desired. In practice, it is not necessary to determine the precise level of coupling. Rather, it is important to try keeping the coupling at a minimal level so that module independence is increased.

Cohesion is a measure of the degree that each module carries out a single, problem-related, and well-understood function. Module strength and binding are other terms that are used for cohesion. There are various types of cohesion; functional and sequential cohesion types are the best, and coincidental cohesion is the worst for maintainability of modules of a system.

Factoring and module size, information hiding, decision splitting, system shape (or morphology), error reporting, editing, restrictivity/generality, and fan-in/fan-out are some of the guidelines that are used as tools to improve an information systems design.

EXERCISES

1. What is the modularity of a system?
2. What are the three major properties of a structured system?
3. What is coupling?
4. What is cohesion?
5. What is the relationship between coupling and cohesion?
6. What are the guidelines to be used to evaluate a module?
7. What are the guidelines to be used to evaluate a group of modules?

8. What are the guidelines to be used for refining a systems design?
9. State the coupling types.
10. What is tramp data?
11. What is bundling?
12. State the cohesion types.
13. Take an available program which has at least four modules. Determine the type of cohesion of each module and the type of coupling of each pair of modules. Use a table to show the type of coupling between the modules. (A module in PL/I is a procedure; it is a subroutine or function subprogram in FORTRAN. In COBOL, program, subprogram, section, and paragraph are examples of modules.)
14. The following modules perform the activities described by the short sentences. Determine the type of cohesion of these modules: G-PAY : Edit time record, get payroll master, and calculate gross pay. WRITE : Write by pen, write by ball-point, write by pencil.
15. Considering the structure chart given as Figure 8.3 in Chapter 8, define the coupling characteristics of modules in a tabular form.
16. Refering to the modules of Figure 12.4, define their cohesion types.

SELECTED REFERENCES

(Jo 80)   Jones, M. P. *The Practical Guide to Structured Systems Design*. Yourdon Press, 1980.

(Pas 72) Parnas, D. L. "On the Criteria Used in Decomposing Systems into Modules," *Comm. of the ACM*, Vol. 15, No. 12 (1972), 1053–58.

(Pes 81) Peters, L. J. Software Design. Yourdon Press, 1981.

(Pr 82)   Pressman, R. *Software Engineering*. McGraw Hill, 1982.

(SMC 74) Stevens, W. P., G. J. Myers, and L. L. Constantine. "Structured Design," *IBM Systems Journal*, Vol. 13, No. 2 (1974), 115–39.

(Yo 81)   Yourdon, Inc. *Structured Analysis/Design Workshop Notes*, 1981.

(YC 79)   Yourdon, E., and L. L. Constantine. *Structured Design*. Prentice-Hall, 1979.

Chapter 13

An Overview of Information Systems Development Methodologies

13.1 A REVIEW OF TERMINOLOGY

As we noted in Section 4.1, algorithm, method, methodology, and strategy are key terms relevant to information systems development efforts.

A dictionary (Web 75) definition of the term *algorithm* is a step-by-step procedure for solving a problem. *Method* is defined as a systematic way, technique, or process of or for doing something; a body of skills and techniques. A body of methods, procedures, working concepts, rules and postulates employed by a science, art, or discipline is known as *methodology*. Finally, *strategy* is defined as the art of devising or employing plans toward a goal.

Although there is no clear distinction between these terms and they are often used interchangeably in the literature (e.g., BG 80, Go 79, Par 78), we would like to use the terms *methodology* and *strategy* synonymously, as well as the terms *method* and *algorithm*.

In engineering if one follows a method (or algorithm) carefully, he/she will surely end up with the correct solution. If, however, one follows a methodology (or strategy), chances of getting one of the possible correct solutions is high, although there is no guarantee for it. Having made this distinction between a method and a methodology, we can now make the following statement: There is not yet a unique, standard method available to be used universally for a successful information systems development; all we have are various methodologies. This clearly indicates that the success of an information systems development process still depends on the skills, experience, and understanding of the systems analyst even if he/she follows an available methodology for information systems development.

13.2 CLASSIFICATION OF AVAILABLE METHODOLOGIES

Almost all of the methodologies that we will be considering have been proposed for the systems design phase of an information systems life cycle. Most of them, however, are also used during the analysis phase of the process. For this reason we will make no distinction between phases in considering and classifying available methodologies. We may need to do so later during the presentation of individual methodologies, however.

The available methodologies for information systems development can be classified into three groups: (1) functional decomposition methodologies, (2) data-oriented methodologies, and (3) prescriptive methodologies. They are further decomposed as follows:

I. Functional Decomposition Methodologies
   1. Top-down approach
   2. Bottom-up approach
   3. HIPO
   4. Stepwise refinement approach
   5. Information-hiding approach
II. Data-Oriented Methodologies
   1. Data Flow Oriented Methodologies
      a. SADT
      b. Composite Design
      c. Structured Design
   2. Data Structure Oriented Methodologies
      a. Jackson's Methodology
      b. Warnier/Orr Methodology
III. Prescriptive Methodologies
   1. Chapin's approach
   2. Design by Objectives (DBO)
   3. Problem Analysis Diagram (PAD)
   4. Problem Statement Language (PSL)/Problem Statement Analysis (PSA)
   5. Others

Functional decomposition methodologies emphasize the relevant dissection of a system into smaller subsystems such that the resulting elementary systems are not so complex to understand, design, and implement. System functions are considered the major concern—therefore, the name "functional approach." The most common examples of functional decomposition methodologies are top-down and bottom-up approaches.

Data-oriented methodologies mainly emphasize the characteristics of the data to be processed. Data flow oriented methodologies and data structure oriented methodologies are two different types of methodologies within that category. Data flow oriented methodologies are generally based on decomposition of a system into modules by considering the type of data elements and their logical behavior within the system. The logical organization of the system is based on the data flow logic and functional relationships between the modules of the system. Data structure oriented methodologies mainly emphasize the output/input data structures of the system. These structures form a base for the structure of the system. The functional relations between the modules or elements of the system and their decompositions are then realized in terms of the system structure.

Prescriptive methodologies are generally computerized procedures to help system development efforts, especially during software system development. The major objective of such a methodology is to free analysts from the detailed technicalities of program design by providing them with a prescriptive approach to analyze the system specifications and to generate the needed software.

In the following chapters these three groups of methodologies are discussed and specific references are cited. The following are some of the general references for the available methodologies: Bergland and Gordon (BG 80), Brookes et al. (BGJL 82), Davis (DaW 83), Freeman and Wasserman (FW 80), Gilbert (Gi 83), Gomaa (Go 79), Griffiths (Gr 78), Leathrum (Lea 83), Parker (Par 78), Peters (Pes 81), Peters and Tripp (PT 78), Pressman (Pr 82), Richardson et al. (RBT 80), Riddle and Fairley (RF 80), Weinberg (Wei 80), and Ziegler (Zi 83).

As a final remark for the available methodologies, one may quote from Riddle and Fairley (RF 80, p. 117):

> . . . any commercial enterprise or government agency that adopts one of these methodologies will do so with a great concerted effort and will make the technique a standard throughout the organization. To do otherwise would not let the technique take hold in the organization, and would allow programmers and managers to fall back on their old methods. Once a method has been adopted by an organization and found to be beneficial, there will be extremely strong resistance to change, no matter how many faults may appear with the methodology. This situation is similar to the adoption of certain programming languages by organizations in the past and has been one of the major deterrents to acceptance of new and better languages.

SUMMARY

There is often no distinction between the terms *algorithm*, *method*, *methodology*, and *strategy*. Although the terms are often used interchangeably, one can group them into two as method (and algorithm) and methodology (and strategy). Normally if one has a method and follows it very carefully, he/she will surely end up with the correct solution. By contrast, following a methodology will most likely lead to one of the possible correct solutions, although there is no guarantee for it.

There is no one single method available to guarantee the successful development of an information system; all we have are various methodologies. This is why the success of an information systems development process still depends on the skills, experience, and understanding of the systems analyst even if he/she follows one of the available methodologies.

The available methodologies for information systems development may be classified as functional decomposition methodologies, data-oriented methodologies, and prescriptive methodologies.

EXERCISES

1. What is the basic difference between a method and a methodology?
2. How are available methodologies presently being classified?

SELECTED REFERENCES

(BG 80)   Bergland, G. D., and R. D. Gordon. *Tutorial: Software Design Strategies.* IEEE Computer Society, 1981.

(BGJL 82)   Brookes, C. H. P., P. J. Grouse, D. R. Jeffery, and M. J. Lawrence. *Information Systems Design.* Prentice-Hall, 1982.

(DaW 83)   Davis, W. S. *Systems Analysis and Design.* Addison-Wesley, 1983.

(FW 80)   Freeman, P., and A. I. Wasserman. *Tutorial: Software Design Strategies.* IEEE Computer Society, 1980.

(Gi 83)   Gilbert, P. *Software Design and Development.* Science Research Associates (SRA), 1983.

(Go 79)   Gomaa, H. "A Comparison of Software Engineering Methods for System Design," *Proc. of National Electronics Conference,* Chicago, October 1979, 464–69.

(Gr 78)   Griffiths, S. N. "Design Methodologies," in *Infotech: Structured Analysis and Design,* Vol. 2, Infotech International, Ltd. Maidenhead Berkshire, UK. 1978, 133–66.

(Lea 83)   Leathrum, J. F., *Foundations of Software Design.* Reston/Prentice-Hall, 1983.

(Par 78)   Parker, J. "A Comparison of Design Methodologies," *ACM SIGSOFT,* Software Eng. Notes, Vol. 3, No. 4 (October 1978), 12–19.

(Pes 81)   Peters, L. J. *Software Design.* Yourdon Press, 1981.

(PT 78)   Peters, L. J., and L. L. Tripp. "Some Limitations of Current Design Methods," in *Infotech: Structured Analysis and Design,* Vol. 2 (1978), 249–64.

(Pr 82)   Pressman, R. *Software Engineering*. McGraw Hill, 1982.

(RBT 80)   Richardson, G. L., C. W. Butler, and J. D. Tomlinson. *Structured Program Design*. Petrocelli, 1980.

(RF 80)   Riddle, W. E., and R. E. Fairley. *Software Development Tools*. Springer Verlag, 1980.

(Web 75)   Webster's New Collegiate Dictionary, G. & C. Merriam, 1975.

(Wei 80)   Weinberg, V. *Structured Analysis*. Prentice-Hall, 1980.

(Zi 83)   Ziegler, C. A. *Programming Systems Methodologies*. Prentice-Hall, 1983.

Chapter 14

Functional Decomposition Methodologies

14.1 INTRODUCTION

As stated earlier, some of the available methodologies such as the top-down approach, the bottom-up approach, HIPO, SR (Stepwise Refinement), and information hiding are classified as *functional decomposition methodologies*. The following sections discuss these methodologies very briefly.

14.2 THE TOP-DOWN APPROACH

The top-down approach is also called "decision analysis" when it is applied to the analysis phase of the systems development life cycle; it is, however, the best known method for systems design. Some other methodologies such as HIPO and SR are special applications of the top-down approach.

In the top-down approach, the highest level decisions of the system are made first and lower level decisions are handled later. If we think of a tree of data structures, the upper branches and leaves (i.e. upper level modules) would be designed and implemented first.

14.3 THE BOTTOM-UP APPROACH

The bottom-up approach is also known as "data analysis" when it is applied to the analysis phase; it is a classical approach in systems design.

In the bottom-up approach one makes the lower level decisions first and higher level decisions are gradually handled later. Again, considering the system structure as a tree in data structures, in the bottom-up approach we design the lower leaves and branches, first.

14.4 HIPO

As we already noted in Chapter 5, HIPO (Hierarchy plus Input-Process-Output) is a documentation tool developed by IBM. Although it is primarily a documentation tool, it is sometimes referred to as a design methodology. Instead of using it as an independent methodology, HIPO is often used to document other methodologies such as composite design or structured design which are discussed later.

14.5 SR (STEPWISE REFINEMENT)

Stepwise Refinement or Iterative Stepwise Refinement (SR or ISR) assumes that an exact and constant problem statement is available and a set of available design alternatives or solutions is presented first. After selecting one of the same level solutions as optimal, we proceed to the next level alternative solutions and repeat the process. Two disadvantages are that there is no rule available when to stop the lower level refinements; also it is difficult to pick the "best" among the presumable equivalent solutions (e.g., Pes 81). Ledgard (Led 73) and Wirth (Wi 71) may be cited as the proper references for SR.

14.6 INFORMATION HIDING

Although information hiding is also referred to as a methodology, it is rather a guideline to be used to improve systems design, as discussed in Section 12.4. The key idea is to specify and design the modules so that the procedures and data contained within a module are not accessible to modules not directly involved with that data. The direct result of this idea is to increase the cohesion of individual modules and minimize the coupling between modules. Recall that these are the major design considerations in systems development. The fundamental concepts of information hiding were proposed by Parnas (Pas 72, Pas 79).

    As a simple example consider a family database. Access to information on the family in the DB should be limited to identifying the family wanted, then accessing information on the individual members.

SUMMARY

Typical functional decomposition methodologies are the top-down approach, the bottom-up approach, HIPO, stepwise refinement, and information hiding. In the top-down approach, the highest level decisions of the system are made first and lower level decisions are handled later. In the bottom-up approach one makes the lower level decisions first and only gradually adds the higher level decisions. In both of these approaches guidelines are lacking on how to conduct the analysis and design of a system, except for some general considerations.

HIPO, which is included as a functional decomposition methodology, is actually a documentation tool instead of a methodology. It is often used to document the top-down approach. It may also be used together with some other methodologies.

Stepwise refinement is an application of the top-down approach. Development of alternative solutions, selection of the "best" alternative, and when to stop the refinement are the major difficulties in applying this approach.

Finally, information hiding is also referred to as a design methodology. Basically, its objective is to develop systems with highly cohesive modules having low coupling between them.

EXERCISES

1. What are the examples of functional decomposition methodologies?

2. What is the basic idea in the top-down approach?

3. What is the basic idea in the bottom-up approach?

4. What are the major characteristics of HIPO?

5. What are major difficulties in applying SR?

6. What is the relationship between information hiding and cohesion/coupling?

SELECTED REFERENCES

(Led 73)   Ledgard, H. "The Case for Structured Programming," *BIT*, Vol. 13 (1973) 45–47.

(Pas 72)   Parnas, D. L. "On the Criteria to be Used in Decomposing Systems into Modules," *Comm. of the ACM*, Vol. 15, No. 12, (December 1972), 1053–58.

(Pas 79)   Parnas, D. L. "Designing Software for Ease of Extension and Contraction," *IEEE Transactions on Software Engineering*, Vol. SE-5, No. 2 (March 1979), 128–37.

(Pes 81)   Peters, L. J. *Software Design*. Yourdon Press, 1981.

(Wi 71)   Wirth, N. "Program Development by Stepwise Refinement," *Comm. of the ACM*, Vol. 14, No. 4 (April 1971), 221–27.

Chapter 15

Data-Oriented Methodologies

15.1 INTRODUCTION

Data-oriented methodologies may be divided into two classes: Data Flow Oriented Methodologies and Data Structure Oriented Methodologies. SADT, Composite Design, and Structured Design are typical examples of Data Flow Oriented Methodologies; Jackson System Design and the Warnier/Orr approach are two typical examples of Data Structure Oriented Methodologies. This chapter is devoted to the discussion of these typical methodologies.

15.2 DATA FLOW ORIENTED METHODOLOGIES

15.2.1 SADT (Structured Analysis and Design Technique)

SADT is a methodology designed by D. T. Ross in the early seventies and supported and developed by SofTech Corporation since 1974. The methodology can be applied to both small and large information systems; there is, however, a relatively limited amount of published information about it (e.g., Co 80, RB 76, RDMcG 77, and Ro 80).

SADT can be applied to both the analysis and design phases of systems development; it is a diagramming technique using the graphical tools of actigrams (activity diagrams) and datagrams (data diagrams) that were presented earlier in

Chapter 8. Each of these diagrams consists of three to six rectangular boxes connected in various ways by horizontal and vertical arrows. Descriptive texts may also be added to the diagrams. For higher level diagrams, descriptive texts may be two to three pages and lower level diagrams should have at most one page of descriptive texts for optimum clarity of diagrams. An example of a SADT activity diagram is given in Figure 15.1, which illustrates major activities, data flows, support mechanisms, and constraints of a hospital information system.

The basic concepts on which SADT is based may be summarized as follows (Pes 81, StB 77):

1. Models are the best means for solving complex problems.
2. Analysis and general design of any information systems problem should be performed in a top-down, modular, hierarchic, and structured fashion.
3. The model used should be graphical to represent the whole structure, its elements, and their interrelationship.
4. The model should represent both happenings (or activities) and things (or data) of the system using two different sets of diagrams.
5. The results of analysis and general design activities must be documented for review, proper feedback, and future maintenance efforts.
6. A disciplined and coordinated team work should be provided in applying the model for analysis and general design of an information system.

The use of "reader-author cycle" in SADT provides the opportunity for disciplined and coordinated teamwork in applying SADT methodology (Figure 15.2). The parties in the cycle are represented by circles, and the communications between them are expressed by directed arcs.

As shown in Figure 15.2, in applying SADT methodology to develop an information system, various people are expected to participate in development efforts. Author, commenter, reader, user, and manager are the titles of some of these individuals. An author is the systems analyst, that is, the individual who applies the SADT model to systems development. A commenter is another systems analyst whose job is to review the SADT models prepared by another systems analyst and to prepare a written comment. A reader is an individual who receives the models for verbal comments or for information only. Users of the system being developed are expected to review the models and submit their written comments on the models. Managers of the cycle are the project manager and users manager(s). For a large system, various other personnel titles may be needed as given in Peters (Pes 81) and Connors (Co 80).

The author prepares an SADT kit—that is, a set of diagrams—and sends them to commenters, readers, users, and managers with a cover letter. Their oral and, in particular, written comments and suggestions for the diagrams are sent back to the author. As the author goes over these comments and suggestions, he/she may agree with some and make needed changes or additions, as well as reexamine certain concepts. Sometimes he/she may need to discuss the comments with the com-

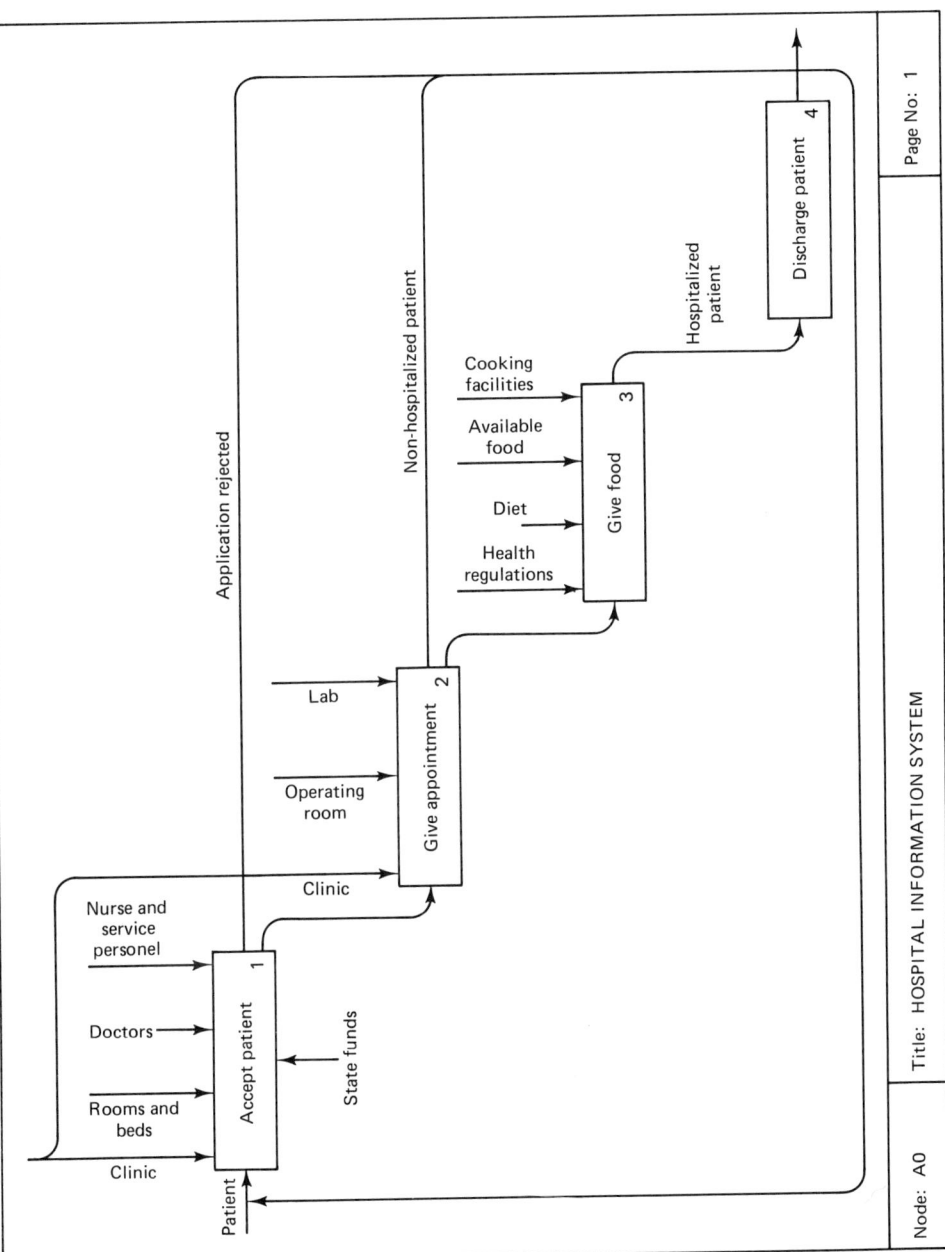

Figure 15.1 Example of an SADT activity diagram

Node: A0   Title: HOSPITAL INFORMATION SYSTEM   Page No: 1

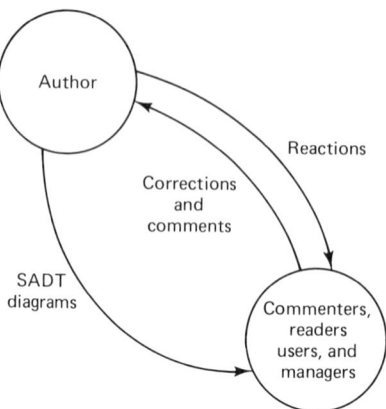

Figure 15.2   Author/reader cycle of SADT

menter, user, or manager. In case a disagreement occurs between the parties, an experienced author or a special committee called TRC (Technical Review Committee) may arbitrate. The author is supposed to prepare written reactions, corrections, and modifications after the comments and suggestions and send them back to commenters, readers, users, and managers once again for additional comments, suggestions, and constructive criticism. That cycle is repeated until the design is finalized and diagrams are put into their final form and printed.

SADT methodology may be summarized as follows: The project team starts with a top-level functional description of the system, that is, a context diagram or activity model. When the author is satisfied that the model represents an accurate view of the required system, the activity boxes of the context diagram are further decomposed at various levels. Remember that activity diagrams represent functional elements and the relevant data elements and data structures of the system. The activity model is only half the job. The author next develops the data model.

Some of the advantages of SADT are as follows:

1. It is teachable.
2. It is an excellent vehicle for communicating with users during the development process.
3. The resulting system design is very well documented as a result of earlier modeling activities.
4. Given the same specifications, most systems designers would end up with the same or similar solutions.

Some of the disadvantages of SADT, on the other hand, are:

1. It may require more time and more personnel for application, resulting in higher costs.

2. The methodology may be used only for the analysis and general (or systems) design phase of the systems development life cycle; for a detailed design, the systems analyst must use another tool or methodology.
3. The processes within the modules are not described.
4. The application of the methodology requires a certain level of skill and experience on the part of the systems analyst.

15.2.2 Composite Design

Composite Design (CD) and Structured Design (SD) were first proposed as software tools to make coding, debugging, and modification easier, faster, and less expensive by reducing complexity. These concepts were later expanded to include information systems development activities. The major ideas of Composite Design and Structured Design belonged to a few people mainly from IBM, L. L. Constantine being the major name among them. The first significant publication on the subject, authored by Stevens, Myers, and Constantine, appeared in 1974 (SMC 74). Although the concepts of Composite Design and Structured Design were quite similar, Myers publicized his approach under the name "Composite Design" (e.g., MyG 73, MyG 75). Myers pointed out the significance of modularity in software design and proposed to use module coupling and module strength (i.e., module cohesion) to obtain modularity.

15.2.3 Structured Design

Although the basic concepts of Structured Design (SD) were identical to Composite Design (CD), SD introduced some additional terminology and concepts such as "transform design" and "transaction centered design," as well as "afferent modules," "efferent modules," and "transform modules." SD also uses the term "module cohesion" for "module strength" of CD. Compared to CD, SD is much more popular and widely used for information systems development activities. There are various publications on SD (e.g., DeM 78, Dic 81, StW 81, YC 79, and Yo 81). Structured Design may be summarized simply as follows:

1. Prepare a DFD (data flow diagram) of the system.
2. Use some or all of the following to develop a structure chart of the system:
   - Transform analysis
   - Transaction analysis
   - Decomposition (factoring or top-down approach)

Thus the objective of SD is to produce a structure chart of a given system using its DFD. One may demonstrate this process in terms of DFD notation as in Figure 15.3 (Yo 81).

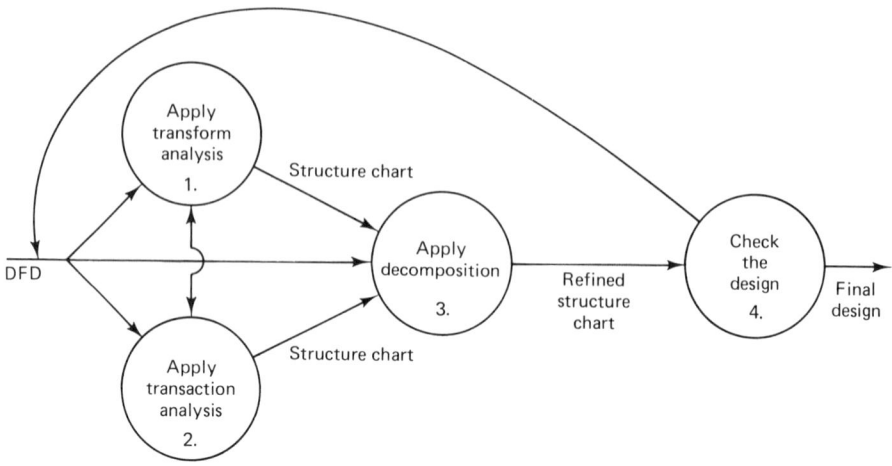

Figure 15.3 Summary of structured design methodology

The process described in Figure 15.3 should be repeated several times until a satisfactory design is obtained.

The basic steps of SD, using transform analysis and decomposition as tools, may be summarized as follows (e.g., Jo 80, Wei 80):

1. Draw a DFD of the system, as illustrated in Figure 15.4a.
2. Identify all of the major input and output data streams in the DFD.
3. Follow each input data stream until it has reached a point at which the stream can no longer be considered input. Identify each such point; the bubbles between such points and source(s) of the system are named afferent (or input) bubbles. They are included in the afferent module of the system.
4. Trace each output data stream backward until it can no longer be considered output. Identify each such point; the bubbles between such points and sink(s) of the system are named efferent (or output) bubbles. They are included in the efferent module of the system.
5. Identify the transformation bubbles (known as "central transforms") in the middle as shown in Figure 15.4b. Transformation bubbles are distinguished from afferent and efferent bubbles by being included in the central transform module.
6. Draw the top two levels of a structure chart; in Figure 15.4c, the function of modules M_C, M_A, M_T and M_E are as follows:
   - M_C is the main module and it acts as a coordinate (or control) module— one that coordinates and manages the activities of other modules.
   - M_A is the afferent module, whose major function is to transform the physical input of the system into logical input; it is therefore sometimes called an input module.

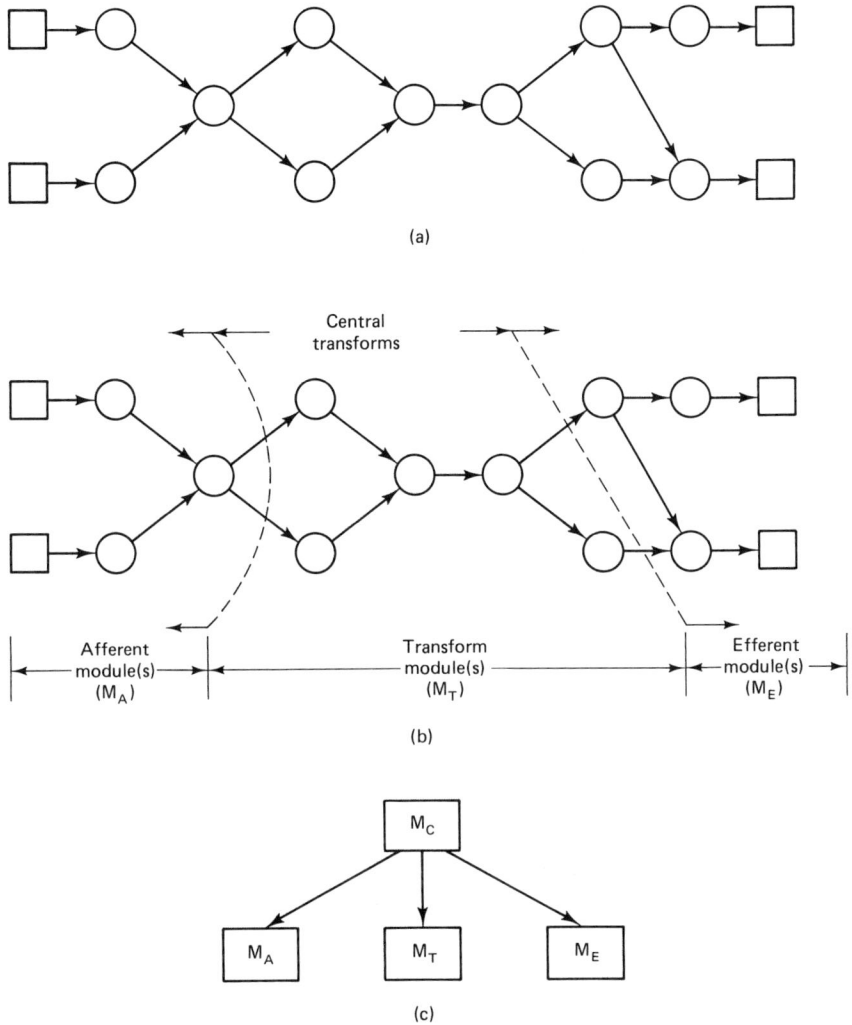

Figure 15.4 (a) A DFD of a system, (b) Identification of the central transforms, and (c) High-level structure chart

- M_E is the efferent module, whose major function is to transform the logical output of a system into physical output; it is therefore sometimes called an output module.
- M_T is the central transform module; it transforms the logical input of a system into logical output.

7. Decompose the second-level modules (i.e, M_A, M_T, M_E) into their components.

8. Refine the decomposition.

An application of the transform analysis is presented in the case study in the Appendix.

Transaction (e.g., Yo 81) may be defined as a data flow that

1. comes in different ''flavors''
2. contains a data element to identify the flavor
3. performs different actions depending on the flavor.

So far, we may summarize the steps of SD using transaction analysis and decomposition as tools as follows (e.g., Wei 80):

1. Draw a DFD of a system, such as the one in Figure 15.5a that shows different transactions and their processes.

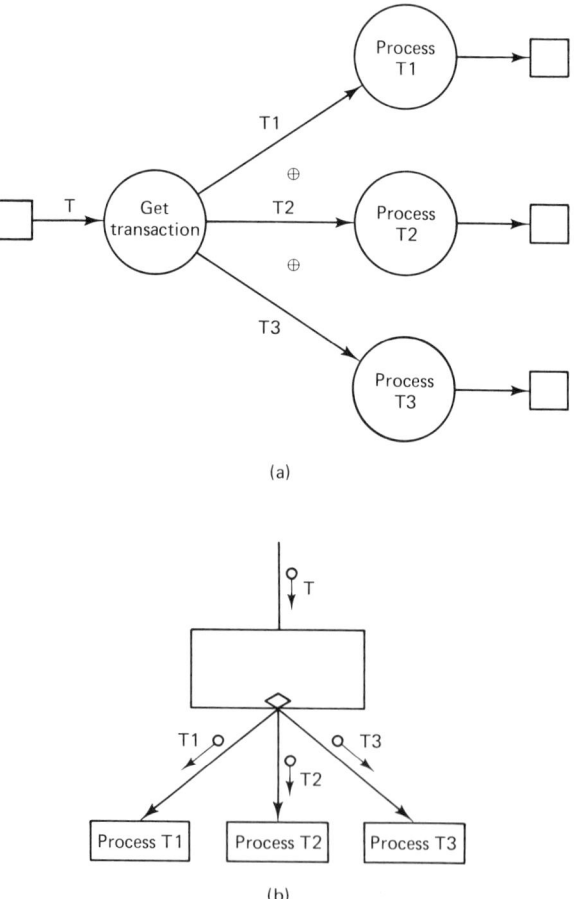

(a)

(b)

Figure 15.5   (a) A DFD of a system, (b) Structure chart

2. Identify any bubbles where an input data stream produces several mutually exclusive data streams by transaction type. Such bubbles are called "transaction centers"; some examples would be addition, deletion, or modification of existing records due to daily transactions.
3. Identify the transactions and their defining actions.
4. Draw a structure chart showing decisions and major transaction modules.
5. Decompose transaction modules.
6. Refine the decomposition.

Some of the advantages of SD are as follows:

1. There are quite a number of publications on the subject and even some companies (e.g., Yourdon Inc., in both the USA and Europe) teaching the methodology.
2. Comparison of various design alternatives is possible because of coupling and cohesion criteria.
3. Resulting structure charts are excellent tools for systems development and for user communication.
4. Although the methodology is not so easy to learn, it is relatively simpler than other available methodologies.

Some disadvantages of SD are:

1. SD does not provide the user with a tool for detail design so another tool must be used.
2. Coupling and cohesion criteria are still qualitative or subjective.
3. The methodology is not so easy to learn.
4. Various users may not end up with the same or similar design results even if they use the same system specifications.

15.3 DATA STRUCTURE ORIENTED METHODOLOGIES

As noted earlier in Chapter 13 and in Section 15.1, the most common methodologies in this group are JSD (Jackson System Development) and W/O (Warnier/Orr or LCS/SSD). Both of these methodologies are discussed in this section and applications are presented in the case study included as an Appendix.

15.3.1 Jackson System Development

Jackson first proposed his approach to software development in 1975 (Ja 75). He extended his ideas for information systems development recently (Ja 83). This subsection is a brief summary of JSD based on M. Jackson (Ja 83, Hic 85).

JSD differs from the other available information systems development methodologies in that its principles combine modelling the real world environment with modelling the functions of a system. JSD is, therefore, more appropriate for developing dynamic systems where the time dimension is a vital factor in contrast to static systems.

As we might expect, the basic terms of JSD such as entity and action have definitions related to time. An entity is defined as an object in the real world which participates in a time-oriented set of actions. An action is an event in which one or more entities participate by performing or suffering the action. The real world is described in terms of entities, actions they perform or suffer, and the orderings of those actions. One may note that the JSD definition of entity is different from that used in a database system where it may be defined as a person, place, thing, event, or concept about which information is recorded.

The development steps of JSD are summarized as follows:

1. Specification Development Phase
   • Specification of the Real World Model
     • Abstract Description of Real World
       a. Entity-Action Step
       b. Entity-Structure Step
     • Realization of the Model
       c. Initial Model Step
   • Specification of System Functions
       d. Function Step
       e. System Timing Step
2. Implementation Phase
       f. Implementation Step

In the first step, the entity-action step, the system analyst describes the real world area of interest by listing the entities and actions with which the system will be concerned. No description of entities is provided at this time. However, for each action, an informal description is given and a list of tentative attributes for that action are provided. The attribute list will be modified and extended in the later development steps.

In the next step, namely, the entity-structure step, the set of actions for each entity is ordered by Jackson Diagram notation that is discussed earlier in Section 11.2. These diagrams, so-called "Structure Diagrams (SD)," together with the list of entities, the list of actions, and their descriptions and attributes, form an integral part of system specification.

The result of the two previous steps is an abstract description of the real world—that is, a model—described in terms of sequential processes. The third step, the initial model step, has the task of stating diagrammatically how the real world process is to be connected to the model process using a new set of symbols given in the Appendix of Jackson's book (Ja 83). The resulting diagram is called a "System Specification Diagram" (SSD). In addition to describing the connection between the real world and the model in this step, the system analyst is supposed to specify the model process by writing a "Structure Text or Meta Code" using the Structure Diagrams of the entity-structure step and considering the System Specification Diagram. Structure Text is a pseudocode that consists of three basic structures: sequence, selection, and iteration.

In the next step, the function step, functions are tied to system outputs, and additional processes are added to the specification as elaborations of Structure Diagrams and/or Structure Text as necessary. The output of the function step is a specification of each of the required functions. This specification is documented using System Specification Diagram or using Structure Text.

The fifth step is the system timing step. In this step the systems analyst considers some potential delays in implementation of different parts of the system and determines, with the system's users, what delays are acceptable for the various parts of the system. The resulting decisions are informally documented, and they provide input to the implementation step; they constrain the implementation alternatives to those which will allow satisfactory response time and satisfactory time lags in the system.

The last step in JSD is the implementation step. In this step, the system analyst considers what hardware or software components are available or will be needed to run the developed system. Documentation of this step is not intended for communication with the system's users, and a new set of symbols will be developed for the "System Implementation Diagram," symbols which are given in the appendix of Jackson's book (Ja 83). JSD structure diagrams are applied in the case study in the Appendix.

Some of the advantages of JSD are:

1. It is a teachable methodology, promoted by two software companies, InfoTech in England and SofTech in USA.
2. It is often the case with JSD that a consensus can be reached on problems.
3. The resulting design is relatively easy to code if the system is a software system.

Some of the disadvantages of JSD are:

1. The heavy emphasis on programming is the major disadvantage; Structure Text that is used for algorithm description is more difficult than pseudocode or Structured English.
2. The system to be considered should be time-dependent; in other words, it should be a dynamic system.
3. Many different symbols in various diagrams (e.g., Structure Diagrams, System Specification Diagrams, System Implementation Diagrams, as well as

the Structure Text) make the methodology too confusing for user communication.

15.3.2 Warnier/Orr Methodology

J. D. Warnier is a French mathematician who has worked in the field of set theory. One of the analytical tools he has developed is a diagram consisting of braces and resembling a horizontal tree structure of a hierarchy chart. The basics of this tool have already been discussed in Chapter 10. The methodology was first proposed as LCP (Logical Construction of Programs) (Wa 74) and later as LCS (Logical Construction of Systems) (Wa 81). The American K. Orr has modified the Warnier Diagrams and applied them to a large class of problems including databases. Orr used the acronyms SSD (Structured Systems Development) (Orr 77) and DSSD (Data Structured Systems Development) (Orr 81b) for his methodologies. Nonetheless, all these methodologies may be referred to as Warnier/Orr or W/O methodologies.

Recall that W/O methodology, similar to JSD, is a data-centered or data-structured design approach, different from functional decomposition and data flow oriented methodologies. W/O methodology enforces the study of a system's outputs as the first step. Once the output of an information system is agreed upon by the system analyst and the user and represented in terms of Warnier diagrams, the data structure of the system is developed next. Once again, the vehicle for both development and documentation is the same type of diagram, only in this case it is used to represent the data structure. A very important property of this methodology is that the resulting design facilitates coding in a software system.

Table 15.1 summarizes the individual steps of some variations on W/O methodology.

Some of the advantages of W/O methodology are:

1. It is teachable.
2. It is relatively simple.
3. It is very convenient for direct coding.
4. The same set of symbols is uniformly used in the development.
5. There is a company teaching the methodology (K. Orr & Associates in the USA).

Some of its disadvantages are:

1. It may get complicated for a large system.
2. Its simple appearance may be misleading.
3. The steps given in Table 16.1 are not that easy to apply for real life system problem.

TABLE 15.1 Summary of W/O Methodologies

Approach Step	LCP	LCS	(SSD) Orr (77)	(SRD) Orr (81)	Higgins (79)	Higgins (83)
1	Define the logical output file	General study (DBs needed)	State the general problem	Logical planning	Define the process outputs	Actual output definition
2	Define the logical input file	Primary data	Identify the structure (and frequency) of the system outputs	Physical planning	Define the logical DB	Logical output definition and design
3	List the logical sequences (flow chart)	Secondary and operational data	Identify the logical database of the system	Logical requirements definition	Define entities and attributes (event analysis)	Logical input design
4	Define the executable operations	Operational output groups	Place the system requirements into a basic system flow hierarchy	Physical requirements definition	Develop the physical database	Logical process design
5	Locate operations in the logical sequence		Check if data required already exists	Logical design	Design the logical process	Physical process—augment
6			Identify events in the real world which affect the DB	Physical design	Design the physical process	Output mapping design
7			Place logical updating actions into basic system hierarchy			Input mapping design
8						Coding
9						Testing

SUMMARY

SADT, Composite Design, and Structured Design are the typical examples of Data Flow Oriented Methodologies. SADT is applied to both the analysis and general design phases of the systems development process. It is a diagramming technique using the graphical tools of actigrams and datagrams. The reader-author cycle is an important tool for the application of SADT methodology. Author, commentater, reader, user, and manager are some of the personnel needed to develop an information systems project. The SADT process involves first preparing activity models and later a data model of the system. Its teachability, its efficiency in user communication, and the good documentation of the resulting system are some of its advantages. Two of its disadvantages are its higher cost and the need for another tool for detailed design.

Composite Design (CD) and Structured Design (SD) were first proposed as software tools. Their use was later extended to information systems development. Basically, Composite Design relies on module coupling and module strength (i.e., module cohesion) as two basic tools to develop modular systems.

Structured Design (SD) is more popular and widely used for information systems development activities. The objective in SD is to produce a structure chart of a system using its DFD. Transform Analysis, Transaction Analysis, and Decomposition are the tools used in SD. Some of the advantages of SD are the relatively high number of available publications, the use of coupling and cohesion for system design measurement or comparison, and its efficiency for user communication. Some of its disadvantages are the facts that it needs another tool for detailed design, that coupling and cohesion are still subjective tools and that it is relatively difficult to learn the methodology.

JSD (Jackson System Development) and Warnier/Orr methodologies are the two methodologies that are grouped under Data Structure Oriented Methodologies. JSD requires that the time dimension of activities of a system or its dynamic quality, be taken into account. Entity and action are the basic terms relevant to JSD. An entity is an object in the real world which participates in a time-ordered set of actions. An action is an event in which one or more entities participate by performing or suffering the action.

The development steps of JSD methodology are entity-action step, entity-structure step, initial model step, function step, system-timing step, and implementation step. JSD is quite teachable and the resulting design is very easy to code if the system is a software system. Heavy emphasis on programming and a plethora of symbols in various diagrams are its disadvantages.

Warnier/Orr methodology is actually comprised of several variations—LCP, LCS, SSD, DSSD, and so on. The output, and user agreement on the output, have the major emphasis in this methodology. Its teachability and simplicity, convenience for coding, and uniformity of symbols are some of the advantages of the methodology. However, the methodology may be complicated for a large information system, and its simple appearance may be misleading. The individual steps of the methodology are neither uniform nor well defined.

EXERCISES

1. What are the methodologies classified as data flow oriented methodologies?
2. What are the major characteristics of SADT?
3. Can you use descriptive texts together with SADT?
4. Comment on the basic concepts on which SADT is based.
5. Describe the author/reader cycle of SADT.
6. Who are the major personnel involved in the author-reader cycle of SADT?
7. Summarize SADT methodology.
8. State some of the advantages of SADT.
9. State some of the disadvantages of SADT.
10. What is the basic idea behind composite design?
11. What are the differences between CD and SD?
12. How can you summarize SD methodology?
13. What are the tools that are used in SD?
14. Describe the steps of Transform Analysis.
15. What is the definition of a coordinate, a transform, an afferent, and an efferent module?
16. What is a transaction?
17. Describe the steps of Transaction Analysis.
18. State some of the advantages of SD.
19. State some of the disadvantages of SD.
20. What are the methodologies that are classified as data structure methodologies.
21. What is the significance of the time dimension in JSD?
22. What is an entity and an action in JSD?
23. Are the definitions of an entity in JSD and in database applications the same?
24. What are the steps of JSD?
25. What types of symbols and diagrams are used in JSD?
26. What is Structure Text? How do you compare it with a pseudocode?
27. State some of the advantages of JSD.
28. State some of the disadvantages of JSD.
29. Who first proposed the W/O methodology?
30. What is the contribution of K. Orr to Warnier's approach?
31. What is the significance of output in W/O methodology?
32. State some of the advantages of W/O methodology.
33. State some of the disadvantages of W/O methodology.

SELECTED REFERENCES

(Co 80)   Connor, M. F. *Structured Analysis and Design Technique.* SofTech May 1980.

(Dic 81)   Dickinson, B. *Developing Structured Systems.* Yourdon Press, 1981.

(DeM 78)   De Marco, T. *Structured Analysis and Systems Specification.* Yourdon Press, 1981.

(Hic 85)   Hiçyilmaz, C. *Jackson System Development Methodology and an Application.* M.S. Thesis, METU, Dept. of Computer Eng., March 1985.

(Hig 79)   Higgins, D. *Program Design and Construction.* Prentice-Hall, 1979.

(Hig 83)   Higgins, D. *Designing Structured Programs.* Prentice-Hall, 1983.

(Ja 75)   Jackson, M. *Principles of Program Design.* Academic Press, 1975.

(Ja 83)   Jackson, M. *System Development.* Prentice-Hall, 1983.

(Jo 80)   Jones, M. P. *The Practical Guide to Structured Systems Design.* Yourdon Press, 1980.

(MyG 73)   Myers, G. J. "Characteristics of Composite Design," *Datamation*, Vol. 19, No. 9 (September 1973), 100–102.

(MyG 75)   Myers, G. J. *Reliable Software Through Composite Design.* Petrocelli, 1975.

(Orr 77)   Orr, K. *Structured Systems Development.* Yourdon Press, 1977.

(Orr 81b)   Orr, K. *Structured Requirements Definition.* K. Orr and Associates, 1981.

(Pes 18)   Peters, L. J. *Software Design.* Yourdon Press, 1981.

(RB 76)   Ross, D. T., and J. W. Brackett. "An Approach to Structured Analysis," *Computer Decisions*, Vol. 8, No. 9 (September 1976), 40–44.

(RDMcG 77)   Ross, D. T., M. E. Dickover, and C. McGowan. *Software Design Using SADT.* Auerbach Publishers, Inc., Portfolio No: 35-05-03, 1977.

(Ro 80)   Ross, D. T. "Structured Analysis (SA): A Language for Communicating Ideas," in *Tutorial on Software Design Techniques*, ed. P. Freeman and A. I. Wasserman. IEEE Computer Society, 1980, 107–25.

(SMC 74)   Stevens, W. P., G. J. Myers, and L. L. Constantine. "Structured Design," *IBM System Journal*, Vol. 13, No. 2 (1974), 115–39.

(StB 77)   Stevens, B. M. "Structured System Design Review," National Bank of Detroit, 1977.

(StW 81)   Stevens, W. P. *Using Structured Design.* Wiley, 1981.

(Wa 74)   Warnier, J. D. *Logical Construction of Programs.* Van Nostrand Reinhold, 1974.

(Wa 81)   Warnier, J. D. Logical Construction of Systems. Van Nostrand Reinhold, 1981.

(Wei 80)   Weinberg, V. *Structured Analysis.* Prentice-Hall, 1980.

(YC 79)   Yourdon, E., and L. L. Constantine. *Structured Design.* Prentice-Hall, 1979.

(Yo 81)   Yourdon, Inc. *Structured Analysis/Design Workshop*, Edition 7.2, July 1981.

Chapter 16

Prescriptive Methodologies

16.1 INTRODUCTION

The methodologies that dictate procedures to be followed during the information systems development process are classified as prescriptive methodologies. Some of these methodologies are in the form of commercially available software packages. The major contribution of such packages is their support for other methodologies by complementing or enhancing the system requirements definition and software development activities of information system development.

In the following sections, some of the commercially available information systems development packages are described. They are ISDOS (PSL/PSA), PLEXSYS, PRIDE, SDM/70, SPECTRUM, and SRES/SREM.

In addition to these software packages, some other methodologies such as Chapin's approach, DBO (Design By Objectives), PAD (Program Analysis Diagram), HOS (Higher Order Software), MSR (Meta Stepwise Refinement), and PDL (Program Design Languages) are classified as prescribed methodologies and they are described by Peters (Pes 81) and Aron (Ar 83).

16.2 ISDOS PROJECT (PSL/PSA)

ISDOS, Information System Design and Optimization System, is a software project being developed at the University of Michigan (e.g., TH 77). The objective of ISDOS is to automate the information systems building process. Its two components are PSL and PSA (Ar 83, Ca 79, and Te 79).

PSL, the major component of ISDOS, is a language for recording user requirements in a machine-readable form. PSL was designed so that its output could be analyzed by the software package PSA. PSL is a language to describe systems, it is not a procedural programming language; PSA is a software package similar to a data dictionary and it is used to check the data as it is entered, to store it, to analyze it, and produce various reports as output. PSA utilizes a DBMS to store the user requirements. PSA analyzes PSL for correct syntax and produces a large number of reports. A data dictionary, a function dictionary, and an analysis of precedence relationships of processes are some of the reports that are produced by PSA. A graphical report is also available; it illustrates all relationships of a given process including whether a process is part of another process or the process has components. PSA performs network analysis to check for the completeness of all relationships between data and processes. It also performs an analysis of time-dependent relationships of the data and an analysis of volume specifications. When the PSA analysis of PSL is error-free, the requirements are passed to system designers to build the system.

As noted by Nunamaker and Konsynski (NK 81), the ISDOS concept was originally intended to automate the system development process; however, very little emphasis has actually been devoted to activities in systems building other than problem definition.

16.3 PLEXSYS PROJECT

The objective of the PLEXSYS Project is to have a system-building system, developed to transform a high-level language problem statement to an executable code for a target hardware configuration. PLEXSYS is really an extension of IS-DOS. While ISDOS has concentrated on nonprocedural aspects of requirements definition, PLEXSYS has concentrated on the automatic code-generation aspects of software systems building.

16.4 PRIDE

PRIDE is an automated system design methodology to be used for information systems development which is offered by the American firm, M. Bryce & Associates. It is described as a thoroughly integrated approach to structured systems analysis/design, data management, project management, and documentation/communications. It provides a Computer Aided Design (CAD) tool for systems development. The software package can actually prepare systems design for manual and/or automated applications. The evaluation of alternatives is just a matter of modifying parameters (BrT 82).

16.5 SDM/70

SDM/70 or Systems Development Methodology/70 is developed and marketed by the American firm, Atlantic Software, Inc. SDM/70 is a comprehensive set of methods, estimation, documentation, and administrative guidelines to help users to develop and maintain effective systems.

16.6 SPECTRUM

SPECTRUM is a systems development methodology developed and marketed by the U.S. firm SII (Spectrum International, Inc.). It has various versions for different purposes such as SPECTRUM-1 (for conventional life cycle), SPECTRUM-2 (the structured project management system), and SPECTRUM-3 (on-line interactive estimator).

16.7 SRES AND SREM

SRES, Software Requirement Engineering System, was developed by TRW for the Software Development System (SDS) of the U.S. Air Force (NK 81).

In SRES, the requirements are stated in the RSL (Requirements Statement Language). REVS (Requirement Engineering and Validation System) is used for analysis of the RSL definition and maintenance of the database. The underlying methodology is called SREM (Software Requirements Engineering Methodology). The system was first implemented on the Texas Instruments Advanced Scientific Computer in 1976. SRES has some concepts similar to those of the ISDOS project discussed earlier.

SUMMARY

Prescriptive methodologies are those that dictate procedures to be followed. Chapin's approach, Design By Objectives (DBO), Program Analysis Diagram (PAD), Higher Order Software (HOS), Meta Stepwise Refinement (MSR), and Program Design Language (PDL) are some of these methodologies.

There are also some commercially available information systems development packages such as PSL/PSA, PLEXSYS, PRIDE, SDM/70, SPECTRUM, and SRES/SREM. The major use of such software packages is to support other methodologies for systems requirement definition and software development activities.

EXERCISES

1. What is the basic property of the prescriptive methodologies?
2. What are the main contributions of prescriptive methodologies in systems development?
3. State the names of some prescriptive methodologies.
4. What is the objective of the ISDOS project?
5. Describe the major components of the ISDOS project.
6. What is the objective of the PLEXSYS project?
7. What is the difference between the ISDOS and PLEXSYS projects?
8. Describe briefly PRIDE, SDM/70, SPECTRUM, and SRES/SREM.

SELECTED REFERENCES

(Ar 83)   Aron, J. D. *The Program Development Process*, Part II. Addison-Wesley, 1983.

(BrT 82) Bryce, T. Private communication.

(Ca 79)   Canning, R. G. "The Production of Better Software," *EDP Analyzer*, Vol. 17, No. 2, (February 1979).

(NK 81)   Nunamaker, J. F., and B. Konsynski. "Formal and Automated Techniques of Systems Analysis and Design," in Cotterman et al., *Systems Analysis and Design*. North Holland, 1981, 291–320.

(Pes 81) Peters, L. J. *Software Design*. Yourdon Press, 1981.

(TH 77)   Teichroew, D., and E. A. Hershey. "PSL/PSA: A Computer Aided Technique for Structured Documentation and Analysis of Information Processing Systems," *IEEE Transactions on Software Engineering*, Vol. SE-3, No.1 (January 1977), 41–48.

(Te 79)   Teichroew, D. "The PSL/PSA Approach to Computer-Aided Analysis and Documentation," Auerbach Publications, Inc., Portfolio No: 32-04-04, 1979.

Chapter 17

Contemporary Research

17.1 INTRODUCTION

The need to single out some measurable properties of programs—for example, their size or complexity—and to use such measures to evaluate and compare the product software has led to the development of "software science." Similarly, there is a growing need to develop some techniques to evaluate and compare information systems quantitatively. Recall that most of the structured tools and methodologies for information systems were first utilized for software development. Similarly most of the measurement and evaluation techniques for software are now being generalized to information systems. In the following sections, we define software science briefly and cite some of the recent research in measurement and/or evaluation of information systems in order to indicate future research activities that might be undertaken.

17.2 SOFTWARE SCIENCE

Software science may be defined as a system of statistically derived formulas that relate measurable properties of computer programs to their length, the time taken to code the programs, and the expected number of errors discovered in the debugging process (HS 80). The late Prof. M. Halstead of Purdue University developed most of these formulas. Some of these statistically derived measures are the vocabulary, program length (or Halstead Length), program volume, program level, language level, programming effort (time), and number of bugs (or program errors during development). We use the following variables to define the various measures:

η_1 = number of unique or distinct operators
η_2 = number of unique or distinct operands
N_1 = total usage of all operators
N_2 = total usage of all operands

The measures are defined as follows (e.g., Gr 84, Ha 78):

1. The vocabulary is

$$\eta = \eta_1 + \eta_2 \qquad (17.1)$$

It is the vocabulary of a specific program under consideration.
2. The program length is

$$N = N_1 + N_2 \qquad (17.2)$$

It represents the length of an expression of an algorithm.
3. The program volume is

$$V = N \log_2 \eta \qquad (17.3)$$

It is a measure of the number of bits required to specify the program.
4. The program level is

$$L \sim \frac{\eta_2}{N_2} \qquad (17.4)$$

It is the ratio of potential volume to actual volume.
5. The language level is

$$\lambda = L^2 V \qquad (17.5)$$

It is a measure of the power of a programming language.
6. The estimated value of programming effort in time is

$$T = \frac{\eta_1 N_2 N \log_2 \eta}{2S\eta_2} \qquad (17.6)$$

Here S is the Stroud number, and it is defined as the number of the most elementary discriminations that the human brain can perform per second.
7. The estimated total number of bugs is

$$B = V/3000 \qquad (17.7)$$

Christensen et al. (CFS 81) have developed the following rules in terms of the above measures:

1. The length of a program is a function of the vocabulary of that program.
2. The lines of code, length, and volume are equally valid as relative measures of program size.

As we shall see in the next section, similar measures may be developed and applied to information systems. Quite a few researchers are working on these objectives at the present time.

17.3 MEASUREMENT OF INFORMATION SYSTEMS

In a study, Troy and Zweben (TZ 81) tried to express the quality of design (in terms of program errors) as a function of 21 independent metrics including coupling, cohesion, complexity, modularity, and size of programs. Examining a set of basic relationships among various software development variables, Basili and Freburger (BF 81) plotted software size against productivity. Similar relations and equations can be used to measure information systems.

Michelson and coworkers (Mic 80) have proposed the following measures to be used for user/system interfaces: reliability, correctness, learnability, usability, flexibility, performance, applicability, security and protection, and cost effectiveness. In a different paper, Cruickshank and Gaffney (CG 80) proposed a set of quantifiable metrics or indicators to measure software design characteristics. Similar to these, Aktas (Ak 82) proposed three groups of metrics relating to the structure, completeness, and resource characteristics of information systems for measurements (Figure 17.1).

In Figure 17.1, the structural characteristics are reliability, flexibility, per-

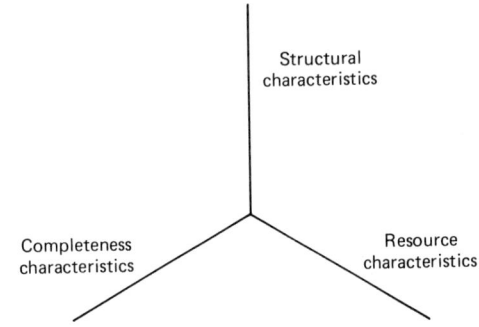

Figure 17.1 Characteristics of an information system

formance, applicability, relative complexity, coupling, cohesion, design logic and presentation clarity, and security and protection. The completeness characteristics are relative requirements, documentation, and completeness; and the resource characteristics are the necessary equipment and cost effectiveness.

Basili and Reiter (BR 81) conducted a small-scale experiment on software development approaches and indicated that a disciplined methodology effectively improved both the process and product of software development. They also verified that methodological discipline has a key influence on the general efficiency of the software development process and significantly reduces the material costs of software development. Similar experiments need to be performed on information systems development projects to quantify procedures and to minimize, if not eliminate, abstract feelings and subjective experiences (Ak 82).

Henry and Kafura (HK 81) defined and validated a set of software metrics which were appropriate for evaluating the structure of large-scale systems, particularly those that are defined for procedure complexity, module complexity, and module coupling.

17.4 OTHER RESEARCH AREAS

Ferrentino (Fe 81) discusses an approach for making good estimates of time and human resources needed to develop a software system. Peercy (Pee 81) described an implemented procedure for evaluating a program's documentation and source code for maintainability characteristics. Harrison and colleagues (HMKK 82) proposed the use of software complexity metrics such as those for program size, data structures and data flow, and program control structures, for program maintenance. Another study on the quality assessment of a software system design is reported by Mohanty (Moh 81). The several metrics used for this purpose are based on the entropy function of communication information theory. These studies also seem applicable to information systems.

Information systems performance evaluation—and various models available for that purpose—is a subject of recent interest as well. In a research paper, Anderson (An 84) reported an application of five performance evaluation methodologies on a relatively small information system.

The problem-solving capability of systems analysts is another interesting area. A relatively recent paper discusses several problem-solving behaviors exhibited by highly regarded and less highly regarded systems analysts and draws some conclusions about the relationship of those behaviors to successful performance (VD 83). Another study has as its subject the behavior of system designers and users in the design and implementation stages (DW 83). Obviously, systems analysts need to understand social and psychological factors for efficient system development.

SUMMARY

Software science may be defined as a system of statistically derived formulas that relate measurable properties of computer programs to their length, the time taken to code the programs, and the expected number of errors discovered in the debugging process. Using the same or similar metrics, various authors have arrived at a number of "quality characteristics" of software and information systems. In addition to information systems metrics, two other topics of recent research are the problem-solving characteristics of systems analysts and user behavior.

EXERCISES

1. Define software science.
2. What are the statistically derived measures for software?
3. What are the metrics to be used to measure some properties of a program?
4. What are the three groups of metrics that can be used to define an information system?
5. What are the results of experiments of disciplined methodologies on software development?
6. State any two recent research topics.

SELECTED REFERENCES

(Ak 82)   Aktas, Z. "Discussion of Structured Systems Analysis and Design Strategies for Information Systems." Paper presented at the ACM Tenth Annual Computer Science Conference, Indianapolis, February 1982.

(An 84)   Anderson, G. E. "The Coordinated Use of Five Performance Evaluation Methodologies," *Comm. ACM*, Vol. 27, No. 2 (February 1984), 119–25.

(BF 81)   Basili, V. R., and K. Freburger. "Programming Measurement and Estimation in the Software Engineering Laboratory," *The Journal of Systems and Software*, Vol. 2, No. 1 (February 1981), 47–57.

(Bo 81)   Boehm, B. W. "An Experiment in Small-Scale Application Software Engineering," *IEEE Transactions on Software Engineering*, Vol. SE-7, No. 5 (September 1981).

(BR 81)   Basili, V. R., and R. W. Reiter. "A Controlled Experiment Quantitatively Comparing Software Development Approaches," *IEEE Transactions on Software Engineering*, Vol. SE-7, No. 3 (May 1981), 299–320.

(CFS 81)   Christensen, K., G. P. Fitsos, and C. P. Smith. "A Perspective on Software Science," *IBM Systems Journal*, Vol. 20, No. 4 (1981), 372–87.

(CG 80)   Cruickshank, R. D., and J. E. Gaffney. "Indicators for Software Design Assessment," Software Cost Eng. Dept., IBM, 1980.

(DW 83)   Dagwell, R., and R. Weber. "System Designers' User Models: A Comparative Study and Methodological Critique," *Comm. ACM*, Vol. 26, No. 11 (1983), 987–97.

(Fe 81)   Ferrentino, A. B. "Making Software Development Estimates Good," *Datamation*, September 1981, 179—82.

(Gr 84)   Gremillon, L. L. "Determinants of Program Repair Maintenance Requirements," *Comm. ACM*, Vol. 27, No. 8 (1984), 826–32.

(HK 81)   Henry, S., and D. Kafura. "Software Structure Metrics Based on Information Flow," *IEEE Transactions on Software Engineering*, Vol. SE-7, No. 5 (September 1981), 510–18.

(HMKK 82) Harrison, W., K. Magel, R. Kluczny, and A. D. Kock. "Applying Software Complexity Metrics to Program Maintenance," *Computer*, September 1982, 65–79.

(Ha 78)   Halstead, M. H. *Elements of Software Science*. North Holland, 1978.

(HS 80)   Halstead, M. H., and V. Schneider. "A Self-Assessment Procedure Dealing with Software Science," *Comm. ACM*, Vol. 23, No. 8 (August 1980), 475–80.

(Mic 80)   Michelson, C. D., et al. "A Methodology for the Objective Evaluation of the User/System Interfaces Using Software Engineering Principles," 18th Southeast Regional ACM Conference, Tallahassee, Florida, March 1980.

(Moh 81)   Mohanty, S. N. "Entropy Metrics for Software Design Evaluation," *The Journal of Systems and Software*, Vol. 2, No. 1 (February 1981), 39–46.

(Pee 81)   Peercy, D. E. "A Software Maintainability Evaluation Methodology," *IEEE Transactions on Software Engineering*, Vol. SE-7, No. 4 (July 1981), 343–51.

(TZ 81)   Troy, D. A., and S. H. Zweben. "Measuring the Quality of Structured Designs," *The Journal of Systems and Software*, Vol. 2 (1981), 113–20.

(VD 83)   Vitalari, N. P., and G. W. Dickson. "Problem Solving for Effective Systems Analysis: An Experimental Exploration," *Comm. ACM*, Vol. 26, No. 11 (1983), 948–56.

Appendix A

A Case Study

A.1 INTRODUCTION

In addition to the examples and exercises provided in the relevant chapters of the textbook, we offer a case study to help clarify the use of some of the tools and methodologies used in an information system development process. Our objective is to demonstrate the application of Structured Design, Warnier/Orr, and Jackson methodologies to the information system of a relatively small, real-life company. In order to ensure privacy and objectivity, we have changed the functions, identities, and locations.

For a better view of the tools and methodologies and for an ease of comparison, the information systems development life cycle that was discussed in Chapter 2 will be followed. The first three phases, namely, planning, analysis, and physical design activities are discussed in sequence.

A.2 CASE COMPANY: THE CAST IRON VALVE COMPANY (CIV CO.)

CIV Company was founded in 1970 as a small workshop with 15 workers. In the following years, because of the high quality of their products, the eagerness of their salespeople and, in particular, the various patent contracts that the company received, both company size and the volume and range of products have increased.

Today CIV Company is working in a new and larger factory complex that houses 250 workers producing various types of valves and fittings for the domestic and international markets. The present organization chart of the CIV Company is given as Figure A.1.

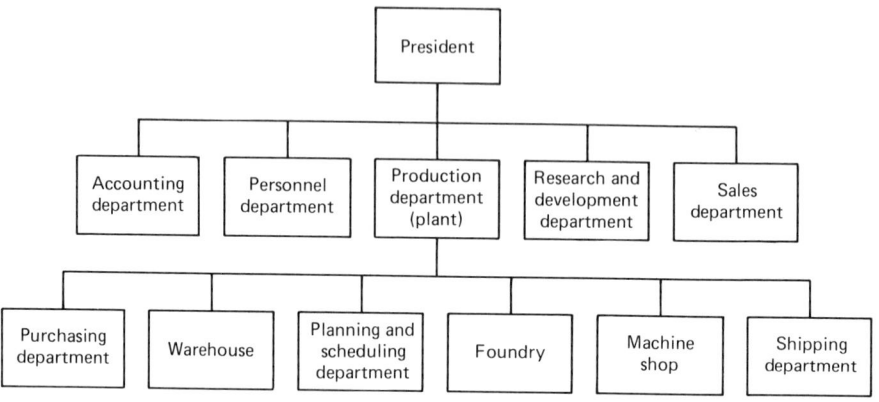

Figure A.1   Present organization chart of the CIV Company

The president of the company, Mr. Schreiber, a talented and a very successful businessman in the region, has long realized the significance of computerization. He is, however, not sure how computers and computerization will be beneficial for his company. For the time being, the most critical problem that bothers him in CIV Company is the delay in meeting customer orders. He is, therefore, anxious to see whether computerization will solve or at least ease this problem.

Mr. Schreiber now hires a system analyst to investigate the feasibility of installing a computer-based inventory control system and to develop and implement it, if feasible. Obviously, if this first application is successful, other applications will follow and eventually a management information system will be realized for the company. However, for the sake of brevity, we are interested only in the inventory control system.

A.3 CASE PROBLEM: INVENTORY CONTROL SUBSYSTEM

Inventory is a subsystem of the information system of a manufacturing organization, along with other subsystems such as sales and invoicing, production and profit planning, accounting, purchasing and accounts payable, and personnel and payroll. An inventory has usable but idle resources—human resources, materials, finished or unfinished machines, monetary assets.

When the resource involved is materials or goods at any stage of completion, the inventory may also be called stock. In many organizations, the ratio of inventory to total assets is estimated to be about 20 to 30 percent. Optimum utilization of inventory is, therefore, vital for a company's profitability, and inventory management and control thus becomes one of the most common applications of information systems.

As noted by Burch and Strater (BS 74), an organization with inventory problems needs an inventory control system to accomplish two things:

1. To make certain that approximately all items are available in the correct quantity when they are needed; that is, to prevent or minimize shortage or stockout. The objective is to maintain sales and customer satisfaction.
2. To prevent an overstock, which is an increase of inventory beyond proper limits. The objective is to reduce investment cost, storage space, and spoilage.

To reach an optimum balance between stockout and overstock situations of inventory, a proper inventory control mechanism must handle the two critical problems of determining the reordering point in replenishing the stock and deciding how much to order, or the value of classical EOQ (Economic Order Quantity). EOQ is the amount of inventory to be ordered at one time for purposes of minimizing annual inventory costs.

The controlling factors in reordering are the lead time (the time interval between placing an order and receiving the product in the firm), the rate of inventory usage, and the safety stock. Safety stock is the level of minimum stock before reordering a particular product to avoid or minimize stockouts and back orders. Back order is a customer order that is not met because of stockout. Some graphical, mathematical, statistical, and forecasting techniques are used to determine the reorder point (e.g., MU 76). The periodic system and the reorder point system are two approaches to replenish inventories. In the periodic system, which is more often used, an order is placed on a specific date. In a reorder point system, an order is placed when the inventory level of an item reaches a predetermined reorder point.

There are various methods of determining how much to order. The commonly used model is the classic EOQ model (e.g., BS 74) which calculates the reorder quantity. In determining that value two opposite types of costs should be considered:

1. Cost for carrying insufficient inventory
2. Cost for carrying too much inventory

The cost elements for carrying insufficient inventory include costly changes in the firm's production rate; extra purchasing, handling and transportation costs; lost sales and penalty costs; and failure in customer relations. On the other hand, the cost of invested capital, handling and storage costs, taxes and insurance cost, deterioration and obsolescence of stocks are some of the cost elements that should be considered in holding or carrying excess inventory. Clearly, then, the objective in determining EOQ is to reach a trade-off that will minimize these costs.

An inventory can simply be modelled as an input/output model such that items that are purchased and/or manufactured and placed in stock are considered input, while those that are sold and/or scrapped are output. The model to be used should also continuously update quantity on hand and make available all sorts of other information.

Although inventory problems may arise in different contexts, finished goods inventory, which is at the hub of the firm's purchasing, production, and sales activities, is one of the most common types of inventory and one that is particularly relevant to the case study at hand. Therefore, in the following sections, inventory will be taken to mean finished goods inventory.

A.4 INFORMATION SYSTEMS DEVELOPMENT PROCESS

The main phases of an information system development process are planning, analysis, physical design, implementation, and maintenance. The individual steps of these phases were already given in Chapter 2 and summarized in Table 2.1 and Figure 2.1. The key considerations and end product of each of these steps are summarized and presented as Table A.1.

The first phase, planning, consists of "study type" activities. The feasibility study is the key activity of this phase, and the resulting report, called a Feasibility Report, will recommend either no change to the existing system, modification of the existing system, or a new system. If a new system is recommended, an outline of the proposed system, proposed requirements, and project schedule will be included in the Systems Proposal which is another end product of Feasibility Study.

The next two phases are the analysis and physical design stages, and the last two are implementation and maintenance. These four phases are all "doing type" of activities compared to the "study type" activities of the planning phase.

The major emphasis of this book and of the case study is on the analysis and physical design phases. We, therefore, assume that the planning phase has already been conducted, that a computer-based system has been shown to be feasible, and that top management has given the go-ahead to begin the analysis phase. After the analysis and general design steps, the detailed design and the next two phases—namely, implementation and maintenance—can be performed in the actual organization. We shall not deal with them in the case study since they are outside the scope of our text.

Section A.5 looks at the case company, Cast Iron Valve Company, from the standpoint of the analysis and design activities for the inventory control subsystem. These phases will be performed according to the model, and wherever there is a need for a structured tool, the three tools—Data Flow Diagram, Jackson Diagram, and Warnier/Orr Diagram—will be applied individually. In the general design step,

TABLE A.1   A Summary of the Information Systems Development Process

Phase	Key Considerations	End Product
I. Planning		
1.1 Request for a system study	•Organization objectives •User department objectives	•Request memo/form to initiate study
1.2 Initial investigation	•Priority of the requested system study •Resource requirements of the requested system •Overall budgeting of the organization/information systems department	•Resource estimate statements •Information Systems Department budget •Statement of objectives and scope
1.3 Feasibility study	•Problem definition •Alternative solutions •Feasibility criteria and cost/benefit analysis •Recommended solution(s)	•Feasibility report •Systems proposal
II. Analysis		
2.1 Redefine the problem	•What is the actual problem?	•Problem definition
2.2 Understand the existing system	•Data gathering •Data analysis	•Data dictionary •General algorithms
2.3 Determine user requirements and constraints on the new system	•Current/future requirements of the new system •Time/resource restrictions •Control points of the system	•Identification of •Outputs/inputs •Operations •Resources
2.4 Logical (or conceptual) model of the solution	•Specification of user needs •Implementation considerations •Training needs	•Logical system design report •Output specifications •Input specifications •Edit, security, and control specifications •Logical data model •Implementation strategy
III. Physical design		
3.1 General design (or systems design)	•How will the system be implemented? •Alternative physical solutions	•General design report •Systems flowchart •List of physical components •Cost/benefit/analysis •Implementation schedule
3.2 Detailed design	•Dividing the work among team members •Report, form, and screen design •Procedures preparation	•The detailed design report •Technical specifications for I/O and storage •Implementation plan •A revised cost/benefit evaluation

TABLE A.1   A Summary of the Information Systems Development Process (*cont.*)

Phase	Key Considerations	End Product
IV. Implementation		
4.1 Systems building	•Planning/scheduling •Hardware/software procurement •Staffing	•The new system
4.2 Systems testing	•Module test •Systems tests	•Test results and feedback
4.3 Installing and conversion	•Conversion strategy	•Changeover plan

4.4 Operations and
       post-implemen-
       tation audits
	•Training
	•Procedures preparation
	•Postimplementation review
		•Trained personnel
		•Procedures
		•Review Report

V. Maintenance		

5.1 Maintenance and
       review
	•System errors
	•Business needs
	•Enhancements
		•Corrected/modified
		  system

the three structured methodologies—Structured Design, Jackson System Development, and Warnier/Orr methodologies—are again considered separately.

A.5 SYSTEM ANALYSIS AND DESIGN OF THE CASE PROBLEM

A.5.1 Analysis Activities

Redefine the problem.   The first step in the analysis phase is to review the problem once more. Here we assume that the president of CIV had already requested a feasibility study of computerization in the company, which had revealed that a computer application would have a significant impact on CIV's customer order processing system. This system is a combination of production, warehousing, sales, and purchasing activities. The major bottleneck, warehousing activities, creates serious dissatisfaction in management, harms customer relations, and causes loss of profits.

The warehouse contains raw materials, parts, and finished goods. The whole inventory system is manually operated. Inventory count, performed twice a year, is neither accurate nor timely. It, therefore, becomes clear that CIV's critical problem is inventory management and control, and the objective of the study is to develop a computer-based-inventory control system.

Understand the existing system.   The system analyst had to understand the existing physical and logical information system for inventory control in order to propose a new logical model at the end of the analysis phase. That model will be used to develop physical design in the phase following analysis.

In order to gain that understanding, the system analyst interviewed Mr. Schreiber, the president, and other managers of the CIV. The analyst held various observation sessions for the activities, in particular inventory flow, of the company and also read some textbooks on the subject of inventory control and management.

The systems analyst collected the existing data and presented them in the following tables and figures of this section.

The existing forms and reports that are relevant to order processing are presented graphically in Figure A.2. As shown in the figure, Customer Order Form (F1) is prepared in the Sales Department and sent to the Production Planning and Scheduling Department. Similarly, the Stock List of Finished Valves Form (F2) and Stock List of Parts Form (F3) are prepared in the Warehouse and sent to the Production Planning and Scheduling Department. The flow of other forms, F4 through F9, are also shown in Figure A.2. Title, description, source, destination, frequency, number of copies, and comments about each of the existing forms (F1–F9) of the order processing system of CIV Co. are listed as Table A.2. In that figure D stands for daily, and M stands for monthly frequency. "Reversible" means that the particular form is sent back to the source from the destination after a certain processing.

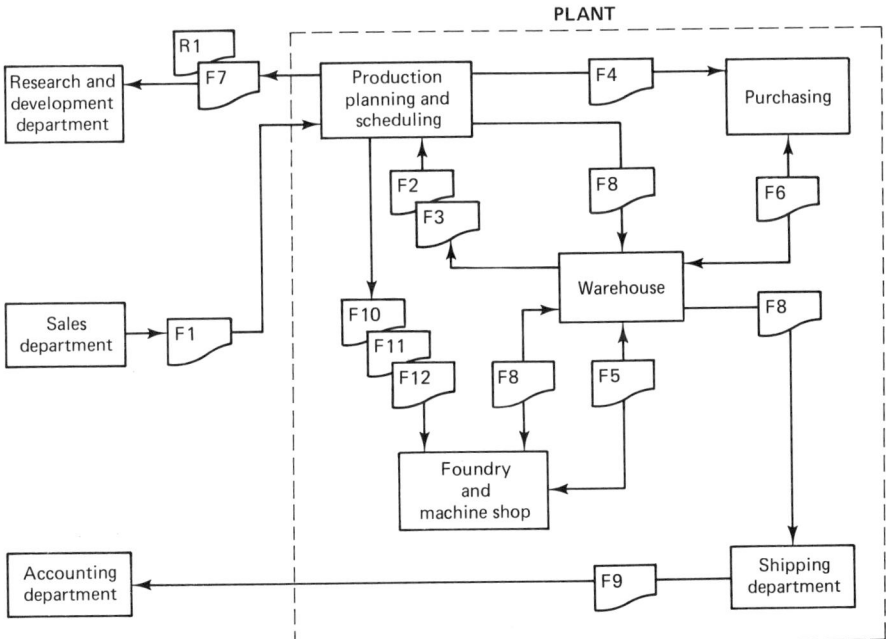

Figure A.2 CIV Company existing order processing form/report flow diagrams

TABLE A.2 CIV Co. Existing Forms Summary Sheet

Title	Description	Source	Destination	Frequency	Number of Copies	Comments
F1	Customer Order Form	Sales	Production Planning and Scheduling	D	1	
F2	Stock List of Finished Valves	Warehouse	Production Planning and Scheduling	M	3	
F3	Stock List of Parts	Warehouse	Production Planning and Scheduling	M	3	
F4	Materials/Parts Purchase Form	Production Planning and Scheduling	Purchasing	D	4	
F5	Warehouse Entry Form for Finished Valves/Parts	Warehouse	Foundry and Machine Shop	D	3	reversible
F6	Warehouse Entry Form for Purchased Material	Warehouse	Purchasing	D	3	reversible
F7	New Product Request Form	Production Planning and Scheduling	Research and Development	D	2	

F8	Warehouse Exit Form	Warehouse	• Shipping • Foundry and Machine Shop • Production Planning and Scheduling	D	3	reversible
F9	Delivery Receipt	Shipping	Accounting	D	2	
F10	Monthly Parts Production Schedule	Production Planning and Scheduling	Foundry and Machine Shop	M	2	
F11	Foundry Work Program	Production Planning and Scheduling	Foundry and Machine Shop	M	2	
F12	Work Shop Work Program	Production Planning and Scheduling	Foundry and Machine Shop	M	2	

(D: daily, M: monthly)

TABLE A.3   CIV Co., Existing Reports Summary Sheet

Title	Description	Source	Destination	Frequency	Comments
R1	New Product Definition and Technical Specifications	Production Planning and Scheduling	Research and Development	M	
R2	Valve Stock Level Summary	Warehouse	Production Planning and Scheduling	Y	
R3	Order Summary	Sales	Production Planning and Scheduling	Y	
R4	Materials Stock Level Summary	Warehouse	Production Planning and Scheduling	Y	
				(M: monthly, Y: yearly)	

Figure A.2 also shows the flow of existing reports of the order processing system. In addition to the New Products Request Form (F7), the New Product Definition and Technical Specifications Report (R1) is sent to the Research and Development Department from the Production Planning and Scheduling Department. The other reports (R2 through R4) are a tabular summary of product stock level, order, and materials stock level of CIV Co. Frequency of these reports are monthly (M) or yearly (Y). Title, description, source, destination, frequency, number of copies, and comments about the reports of the order processing system are summarized as Table A.3. A data element list that summarizes data items in various forms is given in Table A.4.

The existing data items, records, and files are next defined in a manual or a computer-based DD/D (Data Dictionary/Directory) system that is discussed in Chapter 11. Just as an example of the content of DD, the data item "Customer Name" is considered and, using the format of the example in Figure 11.1, it is given as Figure A.3.

Another task in this step of the analysis phase is to present the general algorithms of the existing system. Referring to Figure A.2, we now attempt to summarize the warehouse activities of CIV Co. The existing inventory management system of CIV Co. is given as Figure A.4.

The stock lists produced monthly in the warehouse are not accurate; besides those monthly lists never keep pace with the production scheduling work of the Production Planning and Scheduling Department of CIV Co.

TABLE A.4 CIV Co., Existing Data Items Summary Sheet

Information Systems Development Co.	DATA ELEMENT LIST	Page 15 of 45
Activity No. 003-005	Activity Name: Analysis of the Existing System	Prepared by
Project No. 002-0084-003	Project Name: CIV Co., Order Processing System	Date: 3/15/84

APPEARS IN THE FOLLOWING FORMS

No.	Data Item	F1	F2	F3	F4	F5	F6	F7	F8	F9	F10	F11	F12
1	Date	X	X	X	X		X	X	X	X			
2	Valve Type	X	X	X					X	X		X	X
3	Quantity	X			X			X	X	X	X	X	X
4	Technical specifications	X	X										
5	Delivery date	X			X			X		X			
6	Customer name	X						X		X		X	
7	Salesperson	X								X			
8	Comments		X		X								
9	Part Name			X					X		X	X	X
10	Status Code			X									
11	Material name				X		X		X				
12	Unit				X	X	X		X				
13	Requesting Unit				X								
14	Manager				X								
15	Warehouse Supervisor				X								
16	Plant Director				X								
17	Code No.					X	X		X				
18	Valve/Part Name					X							

175

TABLE A.4 CIV Co., Existing Data Items Summary Sheet (*cont.*)

Information Systems Development Co.	DATA ELEMENT LIST	Page 15 of 45
Activity No. 003-005	Activity Name: Analysis of the Existing System	Prepared by
Project No. 002-0084-003	Project Name: CIV Co., Order Processing System	Date: 3/15/84

		APPEARS IN THE FOLLOWING FORMS											
No.	Data Item	F1	F2	F3	F4	F5	F6	F7	F8	F9	F10	F11	F12
19	Entering Quantity					X	X						
20	Stock Level					X	X		X				
21	Unit Price					X	X		X				
22	Value					X	X		X				
23	Vendor Company						X						
24	Order Date						X						
25	Materials Group						X						
26	Shipment Form No.						X						
27	Transportation						X			X			
28	Product Definition							X					
29	Leaving Quantity							X					
30	Driver's Name									X			
31	Month										X	X	X
32	Operation Code										X		
33	Processing Time in hr.										X		
34	Foundry Unit No.											X	
35	Machine No.												X

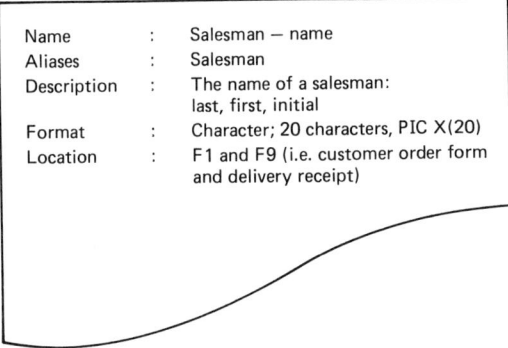

Name	:	Salesman — name
Aliases	:	Salesman
Description	:	The name of a salesman: last, first, initial
Format	:	Character; 20 characters, PIC X(20)
Location	:	F1 and F9 (i.e. customer order form and delivery receipt)

Figure A.3   Example of DD use

At present all the available forms are filed in file cabinets in the warehouse. Clearly, such an application has retrieval and storage area problems.

Determine user requirements and constraints on the new system.
In the existing system, the Purchasing Department, Warehouse, and Planning and Scheduling Department are separate departments. These highly interrelated departments may be combined in a single department, say, Production Control and Management Department, for a relatively small company like CIV Co. The internal operations of the organization, particularly those of purchasing and stocking of materials and finished product, must be subjected to a system of internal control to prevent or minimize waste and fraud.

The new system to be developed should be a modular and hierarchical system; it should also be maintainable, flexible, and testable—in short, a structured system. CIV Co. is a relatively small company. The proposed system should, therefore, be a low-cost system for installation and operation in terms of hardware, software, and personnel. Considering user requirements and constraints on a new system, one may propose some changes in the organization chart as well. The proposed chart, given as Figure A.5, has a significant change compared to Figure A.1 and it should result in greater efficiency.

Logical model of the solution.   As noted earlier, the three commonly used approaches, namely Structured Design, Warnier/Orr, and Jackson System Development, are discussed in the analysis and design phases. In the analysis phase, only the relevant diagrams will be used to represent the logical structure of the proposed system. The relevant diagram for the Structured Design approach in this phase is the Data Flow Diagram. The proposed inventory management system is given in terms of Data Flow Diagrams (Figure A.6), Warnier/Orr Diagrams (Figures A.7 and A.8), and Jackson Diagrams (Figure A.9). Table A.5 is the data flow diagram narrative for CIV's inventory control system.

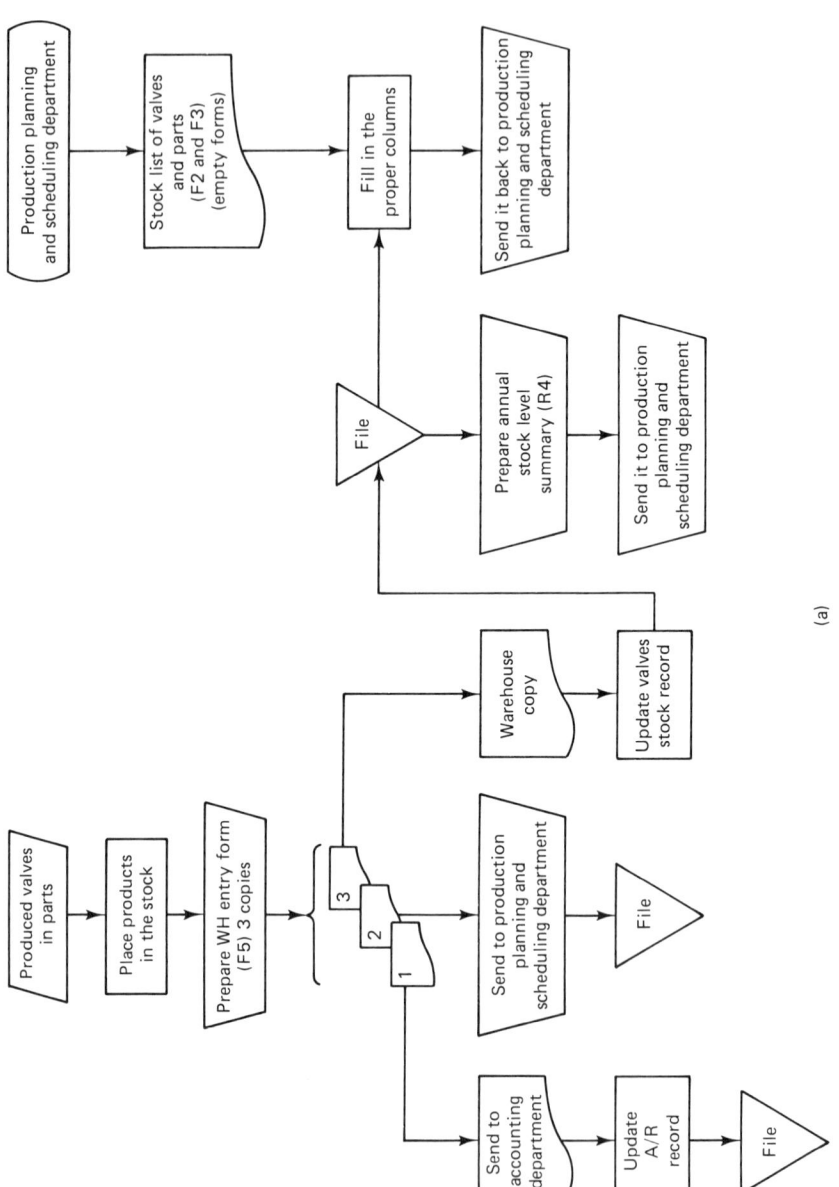

Figure A.4 Existing inventory management system

(a)

178

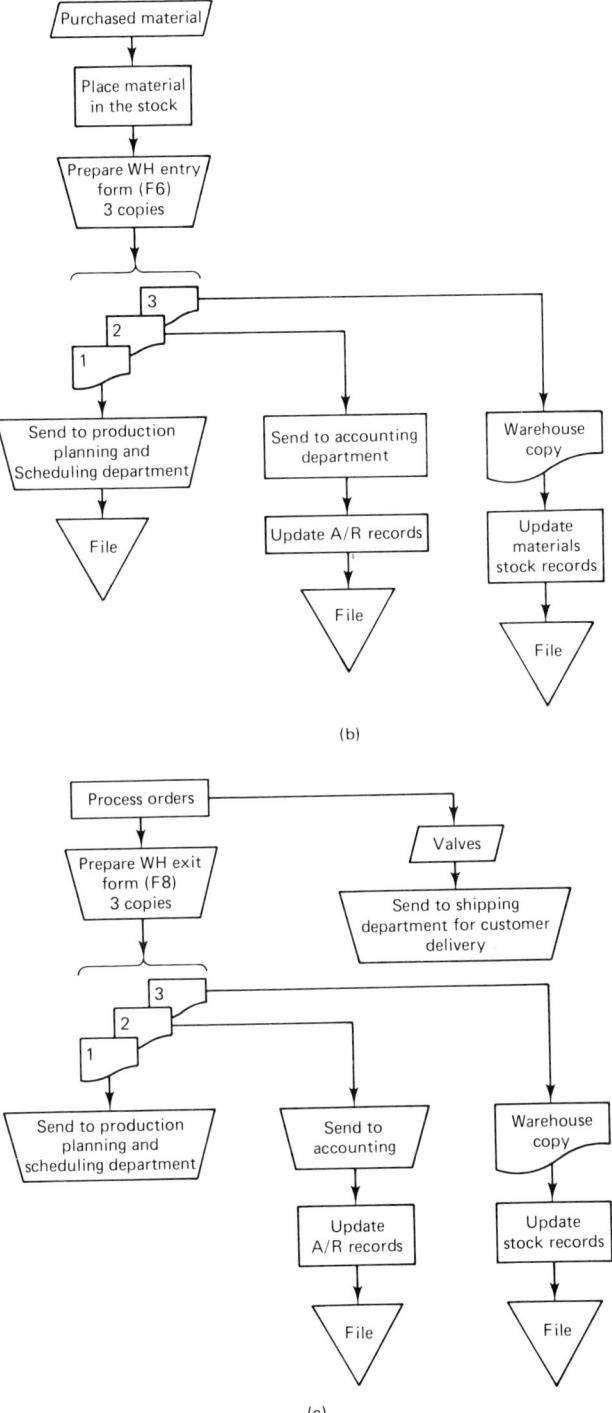

(b)

(c)

Figure A.4 (*cont.*)

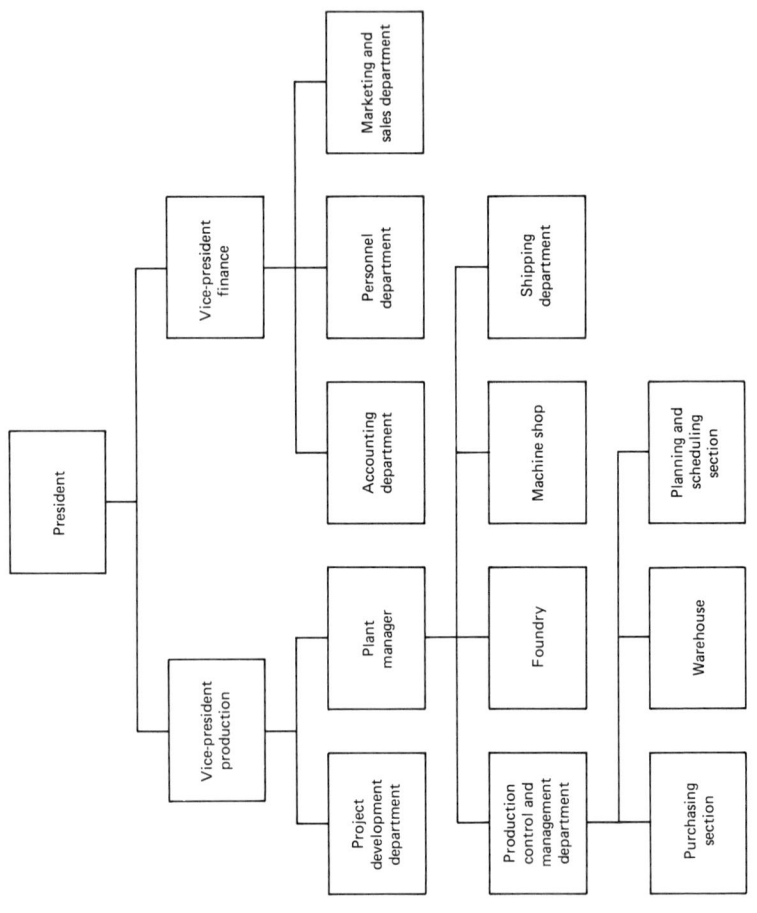

Figure A.5 Proposed organization chart of the CIV Company

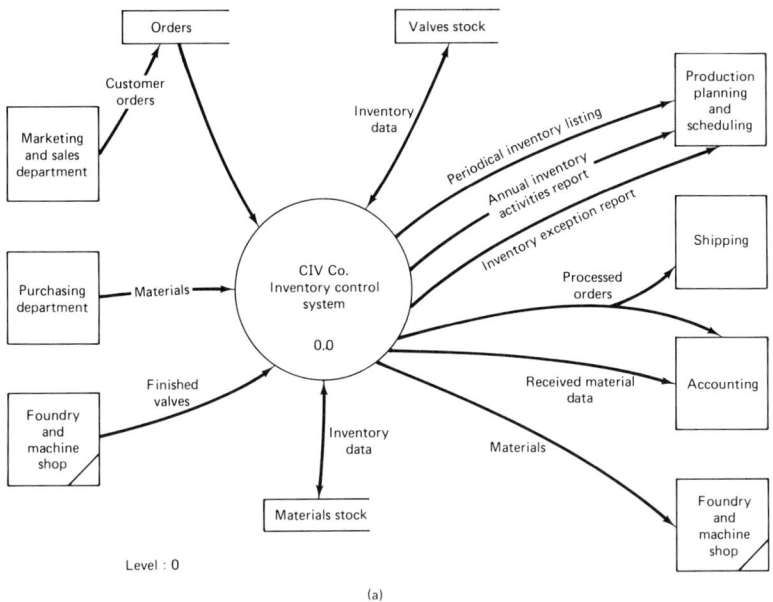

Level : 0

(a)

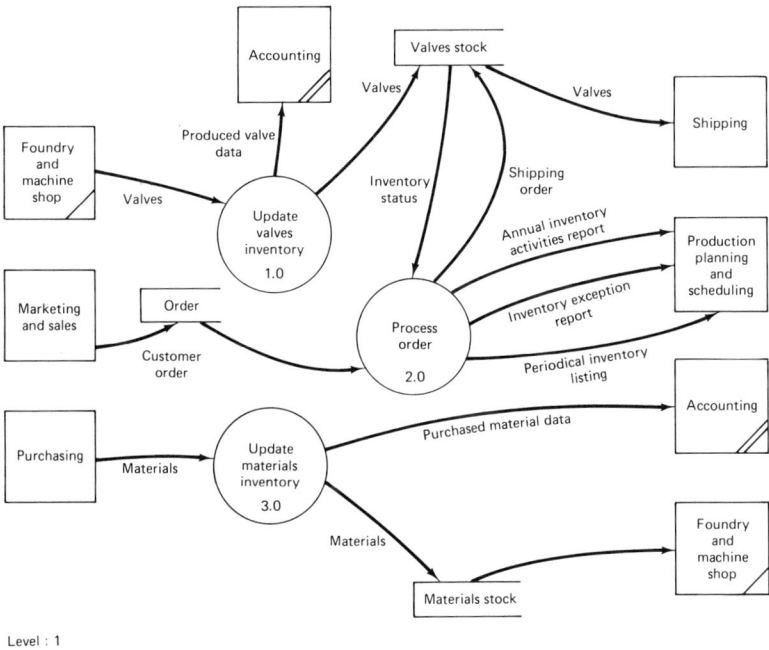

Level : 1

(b)

Figure A.6 (a) Context diagram for CIV Company's inventory control system, (b) Overview diagram for CIV's inventory control system, (c) Process order function (explosion of bubble no. 2) in CIV's inventory control system, (d) Process transaction function (explosion of bubble no. 2.2) in CIV's inventory control system.

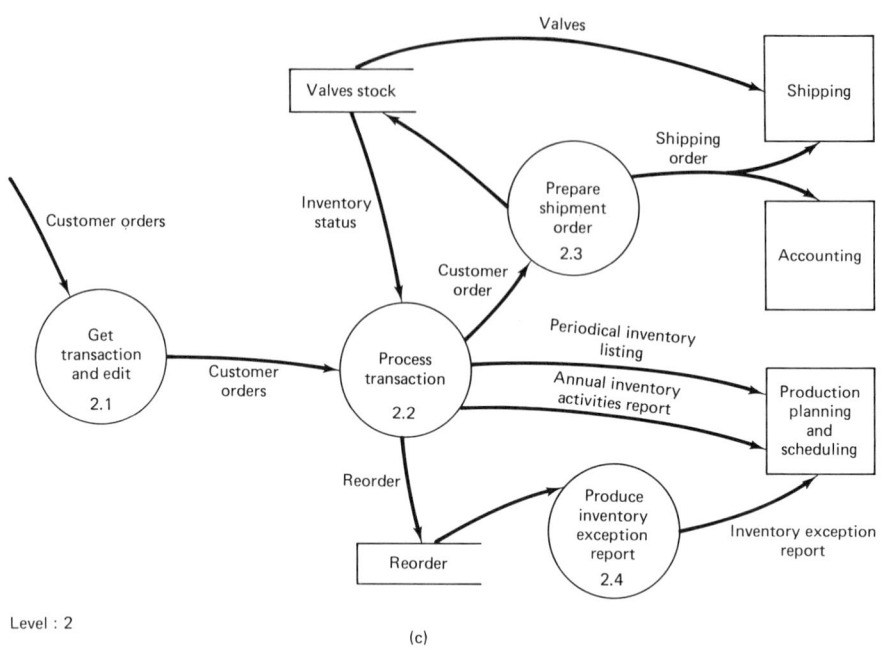

Level : 2

(c)

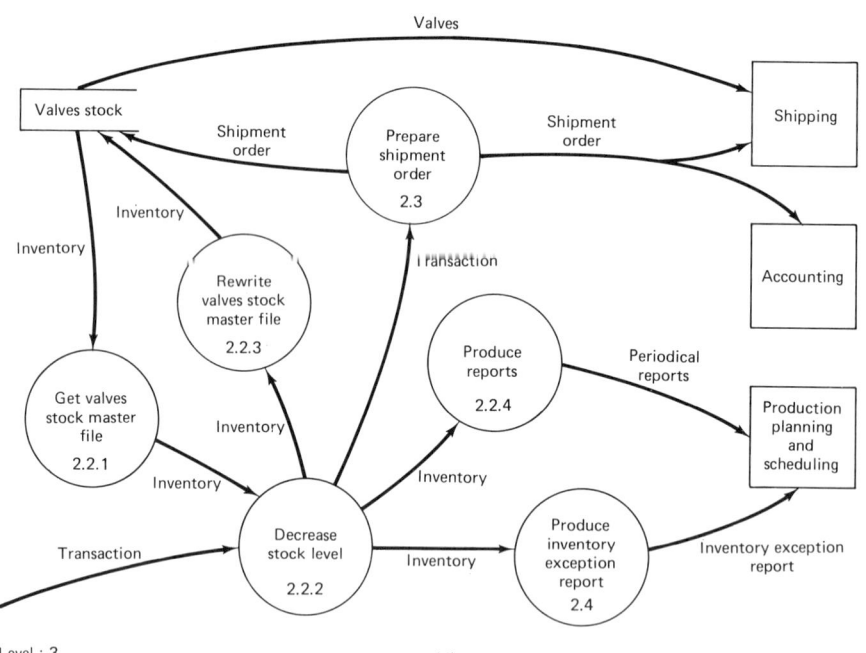

Level : 3

(d)

Figure A.6 (*cont.*)

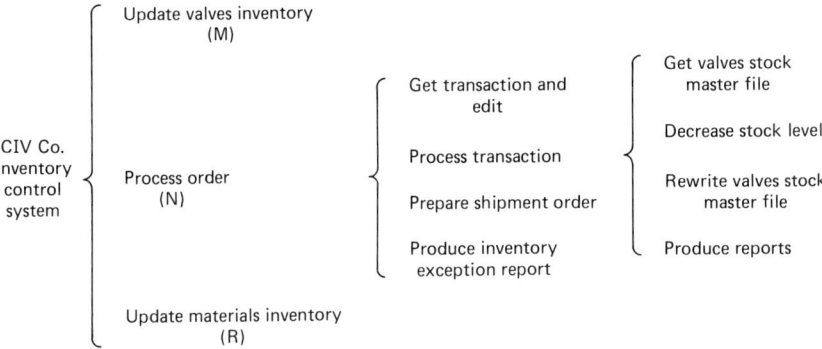

Figure A.7   Warnier/Orr diagram representation of CIV Company's inventory control system

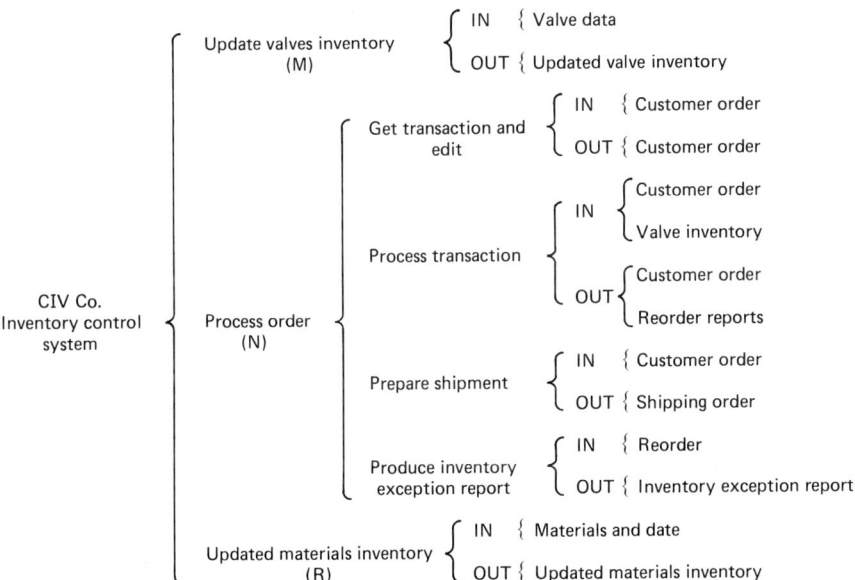

Figure A.8   Warnier/Orr representation of CIV Company's major inputs, outputs, and process of inventory control system

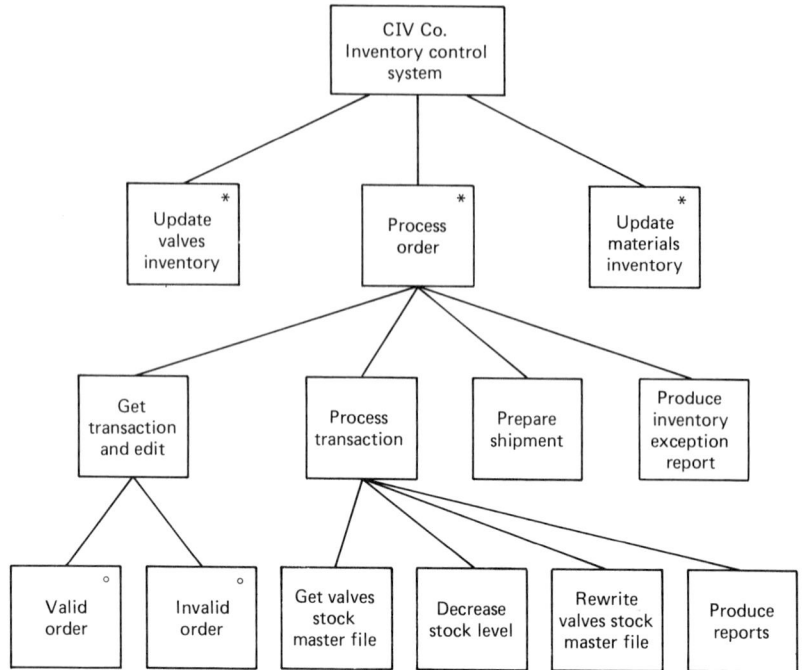

Figure A.9 M. Jackson diagram representation of CIV Company's inventory control system

A.5.2 Physical Design Activities

System design.   The major concern of the system design step of the physical design phase is to study how to implement the new system physically using the logical model proposed in the analysis phase. Various alternative physical solutions should be discussed and one of them should be recommended.

   One alternative is to perform manual operations on the proposed system; another is to develop an information system using a database management system. One could take these alternatives as two extreme cases. A third alternative, which will be recommended, is to use an on-line file management system with sufficient terminals and magnetic disk memory as in Figure A.10.

   Detail design.   The next step in the physical design phase is the detail design step. Our objective is simply to demonstrate how to apply Structured Design, Warnier/Orr, and Jackson System Development; therefore, we are going to look at the detail design of one subfunction, namely, "process transaction" of the Inventory Management System of CIV Co.

   The first application uses Structured Design. Referring to Figure A.6c and

TABLE A.5   Data Flow Diagram Narrative for the Inventory Control System of CIV Co.

1.0	**Update Valves Inventory**	
	Process	Receives produced valves from machine shop and updates valves inventory
	Input	Valves and their data
	Output	Processed valve data for Accounting Dept. and produced valves to stock
	Performance Criteria	Average ten valves per day
	Problems/Comments	During peak production periods, the process is delayed; there is no checking mechanism
2.0	**Process Order**	
	Process	Receives the customer orders, gets the inventory status and processes the order
	Input	Customer Order, Inventory Status
	Output	Shipping Order, Annual Inventory Activities Report, Inventory Exception Report, Periodical Inventory Listing
	Performance Criteria	Daily average 20 customer orders
	Problems/Comments	Manual processing of customer orders is slow; inventory status is not timely; therefore, unfulfilled shipping orders are sometimes possible
2.1	**Get Transaction and Edit**	
	Process	Customer orders are received from Marketing and Sales Department and orders are edited
	Input	Customer Order
	Output	Edited Customer Order
	Performance Criteria	
	Problems/Comments	
2.2	**Process Transaction**	
	.	
	.	
	.	

defining the transform center as in Figure A.11, we end up with the higher level structure chart of Figure A.12. That diagram would further be decomposed and data/control couples would be indicated as we did in Chapter 8.

Similarly, referring to Figure A.7, we rewrite the "process transaction" function of the inventory control system as Figure A.13 in Warnier/Orr Diagram. Later each of the subfunctions shown in Figure A.13 will be decomposed and their elements expressed in terms of W/O Diagrams (see Higgins, Hi 83).

Finally, referring to Figure A.9, we may define the "process transaction" function as in Figure A.14 in terms of Jackson Diagram. Exploding the model further, we can describe each subfunction using Structure Text of Jackson System Design (Ja 83). Remember, Structure Text is very similar to pseudocode and it is used for algorithm description.

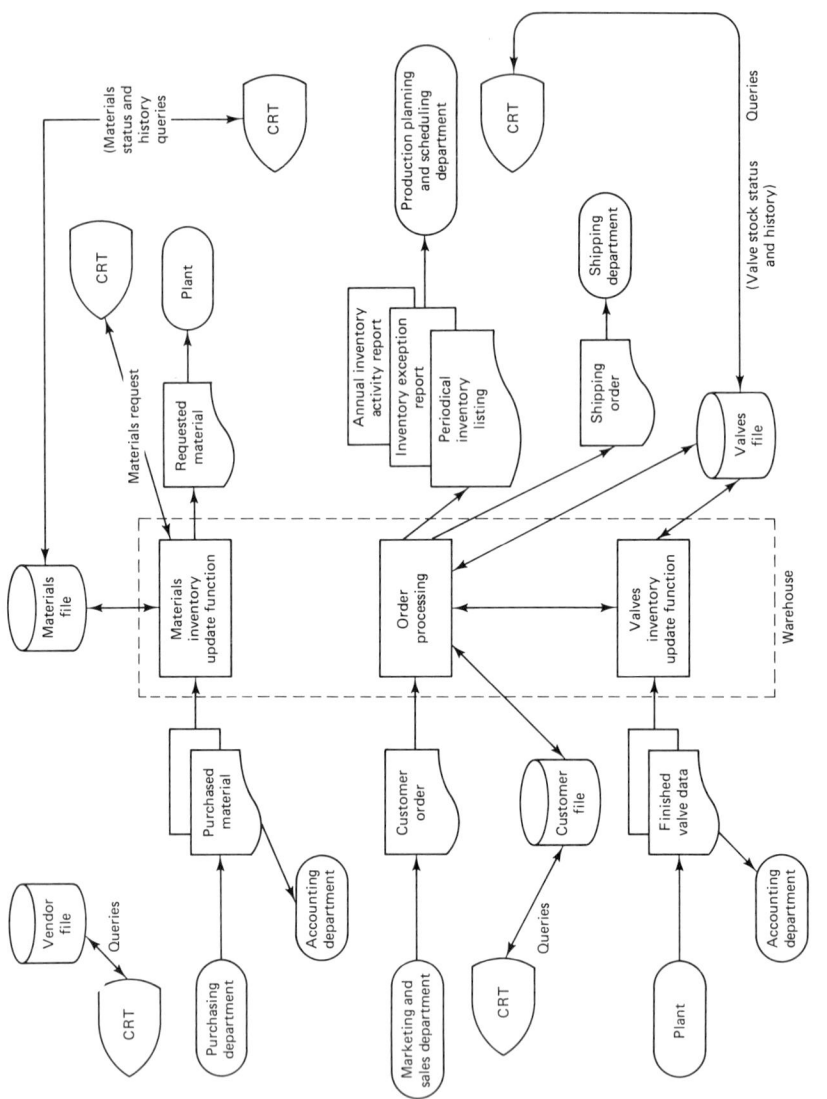

Figure A.10 Recommended system alternative

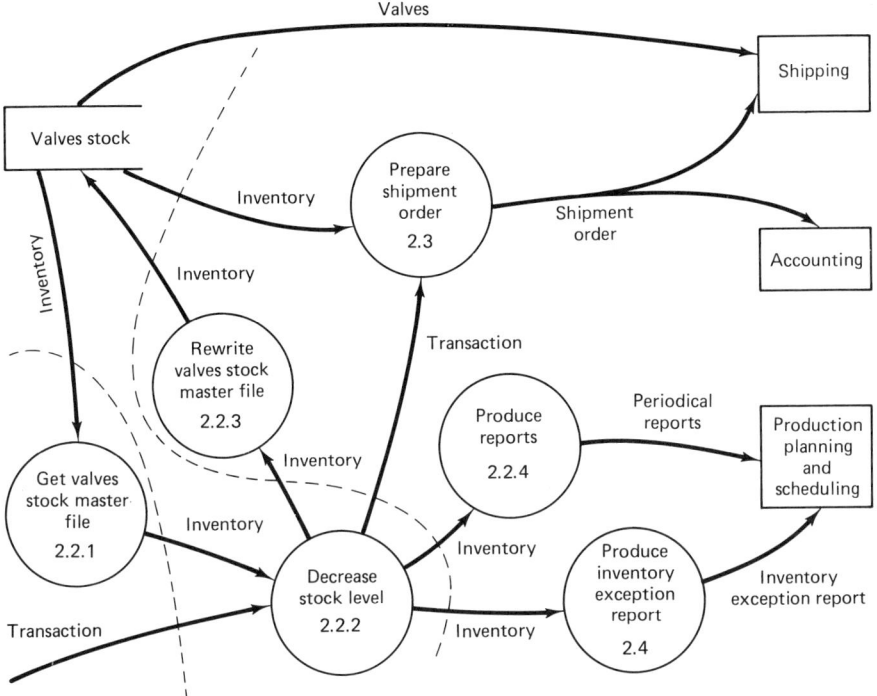

Figure A.11 Main modules of process transaction function

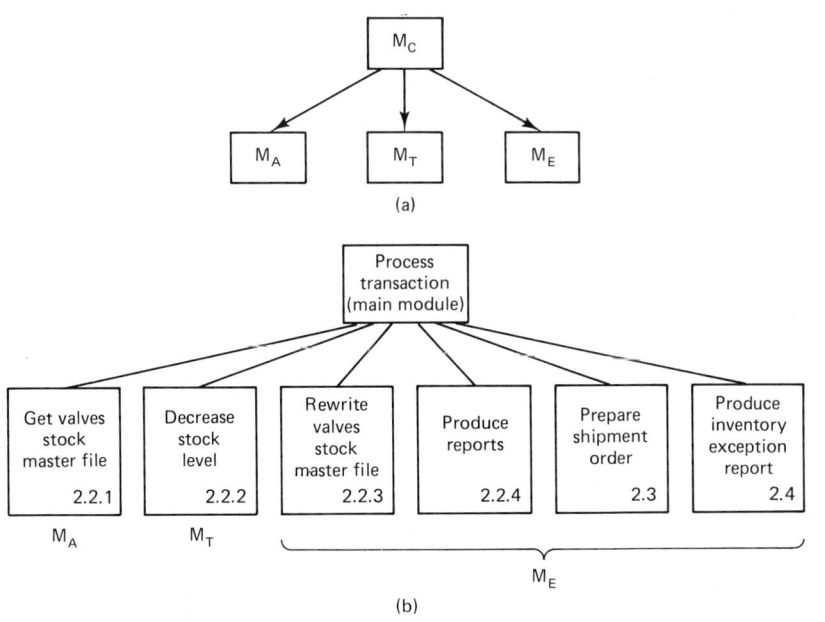

Figure A.12 Structure chart for process transaction module

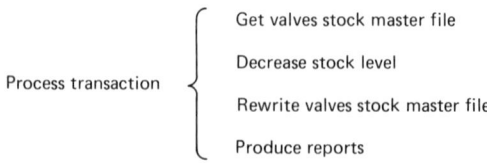

Figure A.13 Process transaction function of inventory control system in Warnier/Orr diagram

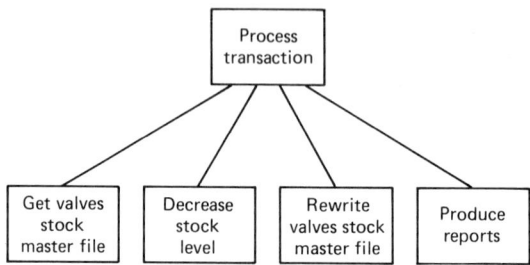

Figure A.14 Process transaction function in Jackson diagram

SELECTED REFERENCES

(BS 74) Burch, J. G., and F. R. Strater. *Information Systems: Theory and Practice.* Wiley, 1974.

(Hi 83) Higgins, D. *Designing Structured Programs.* Prentice-Hall, 1983.

(Ja 83) Jackson, M. *System Development.* Prentice-Hall, 1983.

(MU 76) Matz, A., and M. F. Usry. *Cost Accounting*, 6th ed. Southwestern Publishing Co., 1976.

Index